LLRW Disposal Facility Siting

Technology, Risk, and Society
An International Series in Risk Analysis

VOLUME 8

Editors

Jeryl Mumpower, *State University of New York, Albany, USA*
Ortwin Renn, *Center of Technology Assessment, Baden-Württemberg, Germany*

The titles published in this series are listed at the end of this volume.

LLRW DISPOSAL FACILITY SITING

SUCCESSES AND FAILURES IN SIX COUNTRIES

by

ANNA VARI

Hungarian Academy of Sciences,
Budapest, Hungary

PATRICIA REAGAN-CIRINCIONE

University of Georgia,
Athens, Georgia, U.S.A.

and

JERYL L. MUMPOWER

State University of New York,
Albany, New York, U.S.A.

SPRINGER-SCIENCE+BUSINESS MEDIA, B.V.

A C.I.P. Catalogue record for this book is available from the Library of Congress.

ISBN 978-0-7923-2743-1   ISBN 978-94-011-1120-1 (eBook)
DOI 10.1007/978-94-011-1120-1

Printed on acid-free paper

TABLE OF CONTENTS

Page

LIST OF FIGURES AND TABLES

ACKNOWLEDGMENTS

This book was prepared by the University at Albany, State University of New York, in the course of performing work contracted for and sponsored by the New York State Energy Research and Development Authority (hereafter the "Authority"). The opinions expressed in this report do not necessarily reflect those of the Authority or the State of New York and reference to any specific product, service, process or method does not necessarily constitute an implied or expressed recommendation or endorsement of same. Further, the Authority and the State of New York make no warranties or representations, expressed or implied, as to the fitness for particular purpose, merchantability of any product, apparatus or service or the usefulness, completeness or accuracy of any processes, methods or other information described, disclosed or referred to in this report. The Authority and the State of New York make no representation that the use of any product, apparatus, process, method or other information will not infringe privately owned rights and will assume no liability for damages resulting upon any information contained in this report.

We would like to acknowledge the support of a number of people who helped make possible this book. Thomas H. Attridge in his role as Energy Authority Project Manager was continually helpful and supportive. Kathy Plunkett worked tirelessly to ensure that the production of this volume adhered to high standards, and Lynne Womer and Suzanne Wissel also supported the effort in a number of capable capacities.

Thanks are also due to the more than 30 persons in New York State and the dozens of persons in the other studied states and countries who agreed to be interviewed, providing us with valuable background information. Thanks also are due to the reviewers of this volume, who included Roger Kasperson, Charles Vlek, Felix Oberholzer, Henk Brouwer, and a number of anonymous reviewers. In addition, critical and constructive comments on earlier drafts of this book were submitted by employees of a number of New York State agencies, including the New York State Energy Research and Development Authority, New York State Low-Level Radioactive Waste Siting Commission, New York State Department of Environmental Conservation, New York State Department of Health, and the New York State Energy Office. Likewise, similar comments were provided by employees of a number of public agencies and private companies in the other studied states and countries.

We thank those who have helped to make this book better than it otherwise would have been. Any remaining errors, omissions, or other mistakes are solely our responsibility.

ACRONYMS

AEC	United States Atomic Energy Commission
AECB	Canada Atomic Energy Control Board
AGNEB	Federal Interagency Working Group on Nuclear Waste Management (Switzerland)
ANDRA	Agence Nationale pour la Gestion des Dechets Radioactifs (France)
BEW	Federal Office of Energy (Switzerland)
BLM	United States Bureau of Land Management
BRGM	Bureau des Recherches Geologiques et Minieres (France)
CAC	Citizens' Advisory Committee
CARD	Cortland Against Radioactive Dumping
CCAC	Concerned Citizens of Allegany County
CEA	Commissariat a l'Energie Atomique (France)
CFR	Code of Federal Regulations
CLG	Community Liaison Group
COGEMA	Compagnie Generales des Matieres Nucleaires (France)
CONEG	Coalition of Northeastern Governors
COVRA	Central Organization for Radioactive Waste (the Netherlands)
CSSN	Supreme Council for Nuclear Safety (France)
CVA	Competing Values Approach
DEC	Nebraska Department of Environmental Control
DHS	California Department of Health Services
DEIS	Draft Environmental Impact Statement
DOE	United States Department of Energy
ECN	Energy Research Foundation (the Netherlands)
EDF	Electricite de France
EIR	Environmental Impact Report
EIS	Environmental Impact Statement
EPA	United States Environmental Protection Agency
EQB	Environmental Quality Board
ESA	Environmental Science Associates
EVED	Federal Department of Transport, Communication, and Energy (Switzerland)
GAO	United States General Accounting Office
GIS	Geographic Information System
GTCC	"Greater than Class C"
HALC	Host Area LLRW Commission
HLRW	High-Level Radioactive Waste

HSK	Nuclear Safety Inspectorate of the Federal Office of Energy (Switzerland)
IDNS	Illinois Department of Nuclear Safety
ILRW	Intermediate-Level Radioactive Waste
ISGW Surveys	Illinois State Geological and Water Surveys
KASAM	Special Advisory Committee (Sweden)
KNE	Federal Commission on Nuclear Waste Management (Switzerland)
KSA	Federal Commission for Safety in Nuclear Installations (Switzerland)
LAC	Local Advisory Committee
LB	Legislative Bill
LLRW	Low-Level Radioactive Waste
LLW	Low-Level Waste
LLWD	Low-Level Waste Disposal
LLWIN	Low-Level Waste Information Network
LMC	Local Monitoring Committee
LNNRD	Lower Niobrara Natural Resources Defense
LOFRA	Radioactive Waste Storage Facility Site-Selection Committee (the Netherlands)
MAU	Multi-Attribute Utility
NAGRA	National Co-operative for the Storage of Radioactive Waste (Switzerland)
MDU	Modular Disposal Unit
NACOA	National Advisory Committee on Oceans and Atmosphere
NAS	National Academy of Sciences
NEPA	National Environmental Policy Act
NFS	Nuclear Fuel Services
NGA	National Governors' Association
NRC	United States Nuclear Regulatory Commission
NYCHD	New York City Health Department
NYSDEC	New York State Department of Environmental Conservation
NYSDOH	New York State Department of Health
NYSDOL	New York State Department of Labor
NYSEO	New York State Energy Office
NYSERDA	New York State Energy Research and Development Authority
NYSLLRWAC	New York State Low-Level Radioactive Waste Advisory Committee
NYSLLRWSC	New York State Low-Level Radioactive Waste Siting Commission
NYSLLWG	New York State Low-Level Waste Group

OKG	OKG Aktiebolag (Sweden)
PARD	People Against Radioactive Dumping
RFP	Request for Proposals
SEQR	State Environmental Quality Review
SFR	Swedish Final Repository
SKB	Swedish Nuclear Fuel and Waste Management Company
SKI	Nuclear Power Inspectorate (Sweden)
SKN	National Board for Spent Fuel (Sweden)
SSCAC	Site Selection Citizens' Advisory Committee
SSI	National Institute for Radiation Protection (Sweden)
TLLRWDA	Texas Low-Level Radioactive Waste Disposal Authority
TRCR	Texas Regulations for the Control of Radiation
VLLRW	Very Low-Level Radioactive Waste

FOREWORD

Planning for the management of nuclear wastes -- whatever their level of radioactivity -- is one of the most important environmental problems for all societies that produce utility, industrial, medical, or other radioactive waste products. Attempts to site low-level radioactive waste disposal facilities in Western industrial societies, however, have repeatedly engendered conflicts between governments, encountered vehement opposition on the part of local citizen groups, and given rise to overt hostilities among involved parties. In many instances, the processes have stalled, threatening to incur delays and costs of unknown, but substantial, magnitude.

Many of the most crucial problems about the siting of radioactive waste disposal facilities are not technical, but social. In recent years, acceptability to local communities has played an increasingly large role in the siting process. The experience of certain states and countries in their attempts to site low-level radioactive waste (LLRW) disposal facilities suggests that, while difficult, it is not impossible to achieve public agreement about how best to address this problem.

This book is the result of a study designed to learn more about the causes of failures and successes of efforts to site LLRW disposal facilities. The study is based on case histories of LLRW disposal facility siting processes in six countries. We studied siting processes in five states within the United States (California, Illinois, Nebraska, New York, and Texas) and in five additional countries (Canada, France, the Netherlands, Sweden, and Switzerland). The histories were based on information obtained from public documents and were supplemented by interviews with key participants wherever feasible. The purpose of our effort was to compare and contrast the various efforts to site LLRW disposal facilities and to extract general lessons regarding the factors contributing to their success and failure.

We believe that there are indeed some important lessons to be learned from these case histories. Specifically, we believe that the analyses presented in this book suggest that prospects for success of siting processes are enhanced when (i) the siting process is integrated with broader policy regarding energy production and the nuclear power industry; (ii) the goals of the waste management program are clear; (iii) responsibility for waste management is shared by generators and governments; (iv) responsibilities for site selection, method selection, facility construction, operation, public education, and compensation are assigned to the same organization; (v) site and method selection do not occur simultaneously; (vi) the goal of the site-anu method- selection processes is to identify a licensable site and method with host community support, rather than trying to identify the optimal site or method; (vii) a tailor-made compensation and incentive package is negotiated with the host community, (viii) the host community is directly involved in decision making regarding site selection, method selection, and compensation and incentives; and (ix) the political leaders in both the legislative and executive branches of government display long-term commitment to siting a facility.

We hope that our work helps to document the past and to prepare for the future. Clearly, the problems associated with LLRW management will be important ones for society for the foreseeable future.

Anna Vari
Patricia Reagan-Cirincione
Jeryl L. Mumpower

Albany, New York
August 1993

Chapter 1

INTRODUCTION

During the past four decades, radioactive materials have become increasingly common in industrial societies. Such materials are used for a variety of purposes and processes by the nuclear power industry, hospitals and medical institutions, the pharmaceutical industry, and research institutions, among others. Wastes from the use of these materials are proliferating and accumulating, and managing this waste has become a serious problem throughout most of the industrialized world.

Most industrialized countries differentiate between different types of wastes, at a minimum distinguishing between wastes that are longer-lived and more hazardous (high-level radioactive waste) and wastes that are shorter-lived and less hazardous (low-level radioactive waste). Typically, the former accounts for the vast majority of the radioactivity associated with such wastes; the latter accounts for the vast majority of the volume. Despite the urgency of the problem, siting radioactive waste disposal facilities, whether for high-level or low-level waste, has proved exceedingly difficult in most countries.

This book is concerned with the problem of managing low-level radioactive waste (LLRW), particularly with efforts to site disposal facilities. With only a few exceptions, the story of such efforts is replete with conflict, frustration, delay, inequities, and inefficiencies. In the United States, for example, federal legislation calling for establishment of new LLRW disposal facilities by 1986 was passed in 1980. When the impending failure of that legislation became apparent, it was amended in 1985, extending the deadlines until 1993. Again, the process failed. At present, despite almost a decade and a half of effort involving literally thousands of people, no new facilities have opened, nor is it clear exactly when or where such facilities might open, if ever.

The United States is far from unique. Public controversies about LLRW disposal facility siting, sometimes vehement ones, have arisen in Canada and several Western European countries (Kemp, 1992). In a number of countries, the process has stalled, stuttered, and been delayed. After many years of little or no visible public dissent, opposition to LLRW facilities now appears to be growing in Central and Eastern Europe as well (Novikov, 1993; Vari & Farago, 1991).

Despite ubiquitous problems and obstacles, a few places have succeeded in siting facilities that enjoy broad-based, if not universal, support. The present analysis is based on a set of comparative case histories from six countries. We compare and contrast the various approaches that five states within the United States (California, Illinois, Nebraska, New York, and Texas) and five additional countries (Canada, France, the Netherlands, Sweden, and Switzerland) have used to try to site LLRW disposal facilities.

The selected states and countries are major generators of LLRW, and each has made efforts to establish LLRW disposal facilities during the past decade. They vary widely in the approaches they have adopted to LLRW management, the institutional structures they developed for managing the siting process, the means they used to involve stakeholders and technical experts in the facility siting process, and the amount and type of data they used in making decisions. They also vary considerably in the degree of success they have achieved in their attempts to site facilities. On the basis of our analysis of these ten comparative case histories, we will attempt to extract some general lessons about the advantages, disadvantages, strengths, and weaknesses of the various approaches that have been attempted or implemented.

Before turning to the specifics of these case studies, we discuss very briefly some previous research that is important for understanding LLRW disposal facility siting processes. This includes research about public concerns regarding such facilities, as well as research concerning the key factors that influence the success or failure of siting processes. Next, we discuss the Facility Siting Credo, which represents an important attempt to codify the expertise and accumulated practical experience of researchers and practitioners who have studied or been involved in siting locally undesirable facilities. Then, we describe the Competing Values approach, a theoretical framework that has proven useful for analyzing many complex decision making problems, and which we have used to organize the analysis and interpretation of our ten case studies. Finally, we discuss briefly the study methods that we used and describe the structure of the remainder of this volume.

PUBLIC CONCERNS ABOUT LLRW DISPOSAL FACILITIES

A major factor contributing to the difficulty of siting LLRW disposal facilities has been public concerns about the risks of these facilities (English,

1992; Rosa et al., 1993; Vari, Kemp, & Mumpower, 1991). Research on public concerns has taken two rather distinct approaches. The first focuses on the perceived risks associated with various activities and products. The second emphasizes factors associated with the management and distribution of risks.

The Risk Perception Paradigm

Over the past decade-and-a-half or so, risk perception has evolved from an esoteric, virtually unknown topic that attracted the attention of a mere handful of scholars and practitioners to a high-profile academic specialty in which literally thousands of studies have been conducted. Results from this tradition of research provide some clues as to why public concerns about LLRW have played such an important role in the failure of most countries and states to reach a satisfactory resolution to the problem of LLRW disposal. Although we cannot thoroughly review here the vast literature that has been generated on risk perception, a brief discussion of some of the major concepts and results is helpful.

Drawing on a wide variety of studies, including their own seminal research, Slovic, Fischhoff, and Lichtenstein (1986) proposed a scheme to explain public aversion to some hazards, its indifference to others, and discrepancies between the public's reactions and experts' opinions. They proposed that the general public's risk perceptions can be explained largely in terms of two characteristics or factors (Slovic et al., 1986). The first is the *dread* factor; it is defined by whether hazards are perceived to be dreadful, uncontrollable, involuntary, catastrophic, fatal, inequitable, not easily reduced, and risky to future generations. The second is the *unknown* factor; it is defined by whether hazards are perceived as unknown to those who are exposed, unobservable, having delayed effects, and posing risks not understood by science.

The way that the public typically thinks about risk is different from the way that most technical experts think about it (e.g., Slovic, Fischhoff & Lichtenstein, 1986). Scientists and engineers who work in the field of risk assessment normally define risk in terms of the probability of specific negative consequences, typically mortality, morbidity, or environmental damage. This technical definition of risk has little relation to the factors that the public considers when judging riskiness. Conversely, the factors that the public considers to be most important when evaluating the degree of riskiness of hazards exert little, if any, influence on experts' evaluations.

This means that disputes concerning the risks of siting hazardous facilities are typically not solely a function of different information or factual beliefs. Because the public and technical experts do not think about risk similarly, such disputes are not likely to be resolved simply by better data or more information, contrary to what technically trained personnel often seem to think. These disagreements are often much too fundamental to be resolved so readily; they occur because the various parties have quite different ideas about the meaning of the concept of risk.

In recent years, environmental risk issues have become increasingly salient to the public (Dunlap, 1991). Based on an historical review of controversies involving risky technologies, Edwards and von Winterfeldt (Edwards & von Winterfeldt, 1986; von Winterfeldt & Edwards, 1984) concluded that the most notable change in social values has been the emergence of concerns about the environment and threats to future generations, both of which are central to debates about nuclear waste.

Risk perceptions are influenced by social factors, of course, and the media's influences on public concerns have been a topic of considerable research in recent years. Reviewing much of this research, Kasperson and his colleagues (Kasperson et al., 1988) concluded that perceptions of risk surrounding a hazardous facility are often amplified (or, occasionally, dampened) by the types of information provided to the public by the media or other credible sources such as scientists, special interest groups, and public agencies.

Among the general public, nuclear power and radioactive wastes are among the hazards that are perceived as riskiest and that generate the greatest level of concern, a finding that has replicated cross-culturally in many settings, including Hong-Kong (Keown, 1989), Hungary (Englander et al., 1986), Japan (Kleinhesselink & Rosa, 1991), and Norway (Teigen et al., 1988). On the basis of a series of studies done in Canada, Sweden, Japan, and the United States, Slovic concluded that "nuclear power and nuclear waste continue to be perceived as extremely high in risk and low in benefit to society" (Slovic, 1989; p. 7).

The Risk Management and Distribution Paradigm

Research suggests that, not only do many lay people believe that levels of risk, especially environmental risks, are high and increasing, they also believe that these risks are not being adequately managed or equitably distributed. In a study of hazardous waste facility siting, Kasperson (1986) concluded that public concerns about risk are based largely on distrust of the institutions responsible for risk management. Such distrust stems from perceived past failures of these institutions and perceived inequities in the distribution of risks and benefits among affected parties. In her case studies, English (1992) also observed the presence of the trio of distrust, perceived riskiness, and perceived inequity, although she argued that concern with equity was a result, not a cause of, lack of trust and perception of significant risk.

Kemp (1990) concluded that the nature of concerns that the public expresses about siting LLRW disposal facilities is strongly influenced by contextual factors, especially the nature of formal policy-making and planning processes. These processes create a framework for social discourse that influences the type of concerns expressed by the public. According to Kemp, examples of important contextual factors include, among others, the laws and regulations that pertain to the siting process, the role that independent technical advice plays in the process, and the type of compensation mechanisms that are defined as legitimate for consideration. Kemp's research suggests that, because it influences the types of concerns that are addressed, a siting process can be structured so that it either facilitates success or encourages failure.

Rosa et al. (forthcoming) claim that the birth of nuclear energy technology for the purposes of cataclysmic wartime destruction, the centralized structure of the nuclear industry, and the secrecy surrounding it combined to insulate nuclear development from public participation and public debate. This fostered distrust in and opposition to nuclear energy in general.

On the basis of a review of behavioral research associated with nuclear technologies, van der Pligt (forthcoming) concluded:

Psychological research revealed that local attitudes towards nuclear power stations and nuclear waste facilities are perfectly reasonable from a local perspective. In other words, explaining local resistance in terms of irrational fears is not likely to help solve the issue... Respect for the public's perspective seems essential and a prerequisite for successful policy development.

FACTORS ASSOCIATED WITH SUCCESSFUL SITING PROCESSES

Despite the negative public perceptions associated with many hazardous facilities, especially nuclear ones, facilities are sometimes successfully sited. A number of researchers have identified factors they believe are associated with successful completion of a siting process, based on case studies and analyses of generic siting processes (Williams & Massa, 1983); energy facility siting (Kunreuther & Linnerooth, 1983; O'Hare, Bacow & Sanderson, 1983); hazardous waste facility siting (Davis & Lester, 1988; McGlennon, 1983; Morell, 1984; Ristoratore, 1985); and LLRW facility siting (Armour, 1991; Kemp, 1992; English, 1992). Defining what is meant by the term success is itself no simple matter. This tradition of research generally has defined success as meaning to site a facility that is widely agreed to be appropriate and needed, with public support.

The research has identified four factors that appear to contribute significantly to the prospects for the success of siting processes:

- ◆ Clarity of program goals and the site-selection process;
- ◆ A high degree of public involvement in the process;
- ◆ Continuing political support and leadership; and
- ◆ Development of a compensation or mitigation package for the host community

Each is discussed briefly below.

Clarity of Program Goals and Siting Process

A number of studies have concluded that successful siting processes generally have program goals that are clear and unambiguous. Williams and Massa (1983) concluded that in successful processes, the public understands why a facility is needed and agrees that the facility fulfills societal goals. Morell (1984) argued that the public typically believes there is a true social need for the facility, not just a corporate desire or legislative mandate, in cases where siting approval is obtained. In a study of radioactive waste management, Kemp (1989) concluded that successfully siting a facility critically depends on clarity about the purpose and direction of the overall management policy, as well as on the merits of the individual case.

A clearly designed site-selection process also appears to be important to the success of the process. Davis and Lester (1988) propose that statutes should provide structural details concerning how alternatives will be rated or ranked, how financial resources will be allocated, and how to integrate efforts among implementing agencies. Applicable laws and regulations should specify or recommend decision rules, encourage public officials to participate in the siting process, and provide for formal access to the process by outsiders (Davis & Lester, 1988). Both McGlennon (1983) and English (1992) concluded that, because the site-selection process should be publicly defensible and legitimate, success is more likely if the design of the process is open to review by interested parties.

Public Involvement

A number of studies have concluded that successful siting processes tend to be accompanied by a high degree of public involvement. According to

McGlennon (1983), successful processes typically address concerns about risks through a public procedure that includes informational sessions, seminars on technology, and meetings with local government officials. Renn (1993) argued that in order to establish a successful siting process, conditions for competent and fair public participation have to be assured; criteria for competence and fairness are discussed in detail by Renn et al. (forthcoming).

In successful processes, environmental and health risks are typically clearly delineated and publicly discussed (Williams & Massa, 1983). Open communication by government agencies may prove helpful in overcoming public distrust and establishing the agencies' integrity and credibility (Morell, 1984; English, 1992).

Clear, direct, and honest communication encourages public participation and improves the chances of success. Technical information should be translated into language that is easily understood by the public (Williams & Massa, 1983) and packaged to meet the varying needs of interested parties (O'Hare, Bacow & Sanderson, 1983).

Several researchers have concluded that the prospects for success are also improved if interested parties are empowered to make choices regarding site selection and facility design (Armour, 1991; Massam, 1993). Siting processes in which a single candidate site or facility design is proposed without any prior public involvement or knowledge of the selection criteria often result in failure (Ristoratore, 1985).

Continuing Political Support and Leadership

Davis and Lester (1988) concluded that successful siting processes usually enjoy continuing political support of statutory objectives and committed political leadership. State governments demonstrate their commitment by providing the time and financial resources to carry out the process adequately (McGlennon, 1983). State-level commitment to the process by itself is often insufficient, however. Local involvement early in the siting process has been identified as a crucial factor for success (Kemp, 1992; Ristoratore, 1985). Consequently, Morrell (1984) suggested that siting authority should be balanced between state and local agencies. Massam (1993) concluded that responsibility for siting decisions should be shared among various stakeholders, including state and local governments, technological experts and lay citizens, with clear lines of accountability.

Mitigation and Compensation Packages

Kunreuther & Linnerooth (1983) concluded that local political support is ordinarily encouraged by mitigation or compensation packages. In successful processes, people in the potential host community believe the local benefits outweigh the local costs (McGlennon, 1983). According to Ristoratore (1985), compensation packages need to be custom-designed for a community to be effective. Morell (1984) argued that it is important for them to be developed through a process of negotiation. Several researchers have concluded that non-financial incentives, including the opportunity for independent monitoring and some degree of control over the operation of the facility, may promote public acceptance more strongly than financial incentives (Carnes et al., 1982; van der Pligt, forthcoming).

Local support is further solidified if contractual agreements between the community and the institutions in charge of siting the facility are explicit (Kunreuther & Linnerooth, 1983). Armour (1991) argued, however, that no compensation package is likely to prove successful unless it is clear that stringent safety standards are satisfied.

THE FACILITY SITING CREDO

A substantial amount of experience about the siting of facilities has been generated. Much of this experience has not been analyzed formally or published in widely available outlets, but is nonetheless potentially extremely valuable. Recently, an effort was made to use the expertise and accumulated experience of researchers and practitioners to develop a set of guidelines for institutions charged with siting locally unwanted land uses (LULU's).

In Autumn 1989 and Winter 1990, siting experts from across the country met during a two-part National Workshop on Facility Siting, sponsored by the MIT Hazardous Substances Management Program, the MIT-Harvard Public Disputes Program, and the Wharton Risk and Decision Processes Center at the University of Pennsylvania. This workshop brought together academic researchers, public officials, and representatives of the private sector, all of whom had studied or participated in different siting problems. The attendees included persons familiar with both successful and unsuccessful attempts to site facilities. The workshop reexamined siting theory, tested its limits on specific case studies, and evaluated a new set of principles.

The Workshop produced the Facility Siting Credo, which proposes a set of 13 principles to improve the process by which facilities are sited. The Credo states that institutions charged with siting responsibilities should:

♦ Seek consensus through a broad-based participatory process;
♦ Work to develop trust;
♦ Get agreement that the status quo is unacceptable;
♦ Choose the facility design that best addresses the problem;
♦ Seek acceptable sites through a volunteer process;
♦ Consider a competitive siting process;
♦ Work for geographic fairness;
♦ Keep multiple options on the table at all times;
♦ Guarantee that stringent safety standards will be met;
♦ Fully compensate all negative impacts of a facility;
♦ Make the host community better off;
♦ Use contingent agreements; and
♦ Set realistic timetables.

The Wharton Decision Processes Center is presently conducting additional research on these guidelines to determine the relative importance of different siting principles and identify additional criteria leading to the successful siting of a facility. The study is analyzing questionnaire responses from individuals who have worked to site locally unwanted but regionally necessary facilities across the United States.

Preliminary results (Kunreuther, Aarts & Fitzgerald, 1992) indicate that respondents believe that establishing trust between the developer and host community is a key factor in facilitating the siting process. In the opinion of the respondents, the siting process is most likely to be successful when the community perceives the facility design to be appropriate. The respondents also indicated that public participation is an important variable, especially when it leads toward the view that the facility best meets community needs.

THE COMPETING VALUES APPROACH

The difficulties in siting LULU's have led to the now almost universally familiar NIMBY ("not in my backyard") phenomenon, a problem no less significant for having become perhaps all too familiar. Proposed solutions to the problem are highly diverse.

Few persons, however, have advocated exercising political power through the mechanism of eminent domain to site a facility over the objections of local hosts. Instead, the emphasis is ordinarily on designing an improved decision making process. For instance, Easterling and Kunreuther (1992) propose a highly participative process that they argue will help to reduce perceived risks to health, environment, and economy; provide economic benefits to host communities that will offset negative impacts; convince the public that the proposed facility is appropriate, addresses important needs, and represents the best approach to the problem; assure both procedural and distributional fairness in the procedure used to select sites; and instill an authentic sense of control over the process.

Although their proposal is a thoughtful, highly promising one, its precise details are not the critical issue here. The key point is simply that their proposed improvements to the siting process, like most such proposals, have multiple objectives. Not only are there multiple objectives, different concerned parties may disagree about precisely what the objectives of the process should be and their relative importance.

There is no free lunch when it comes to designing a site-selection process. It is difficult, if not impossible, to design and implement a process that simultaneously satisfies all the desirable attributes that one would ideally want such a process to possess. Many persons have argued, for instance, that siting processes should be participative and flexible. Other persons have argued that the decisions should be accountable and efficient. Some persons have argued that the process should simultaneously be all of these things -- participative and flexible, plus accountable and efficient. Such expectations inevitably contradict or, at a minimum, compete with one another. As more emphasis is placed on public participation, less emphasis can be placed on the efficiency of the decision process. As concern for the accountability of the decision increases, the adaptability of the process inevitably decreases.

When arguments are made, for example, that there is "a need for greater public participation" in siting processes, the important, silent, and sometimes forgotten qualifying phrase "all else remaining equal" needs to be understood. Clearly, a process that greatly increased the role of public participation but lost contact altogether with "the data" would quickly attract criticism from numerous stakeholders, even those promoting greater public participation.

A conceptual framework that helps to clarify the contradictions inherent in facility-siting processes is the *Competing Values* approach to organizational analysis (Quinn, 1988; Quinn & Rohrbaugh, 1981, 1983; Lewin & Minton, 1986). As shown in Figure 1, in the Competing Values theory, two basic value dimensions combine to define perspectives on effective decision making processes. The first value dimension is related to the structure of the process; an emphasis on flexibility competes with an emphasis on control. The second value dimension is related to the focus of the process; an emphasis on the desires of individual stakeholders competes with an emphasis on the needs of the collective. The relative emphasis on these competing values defines four distinct perspectives on effective decision making: the *rational* perspective, the *empirical* perspective, the *consensual* perspective, and the *political* perspective (Rohrbaugh, 1987).

A third key consideration in evaluating decision processes is whether emphasis is placed on process (means) or outcomes (ends). Within each perspective, processes and outcomes are linked. For example, in the consensual perspective, an emphasis on participatory processes is linked with a concern for supportable decisions. Combining the four perspectives with the distinction between means and ends results in the eight criteria describing decision processes (Quinn, Rohrbaugh, & McGrath, 1985), as indicated in Figure 1.

Figure 1. Framework for Competing Values Approach.

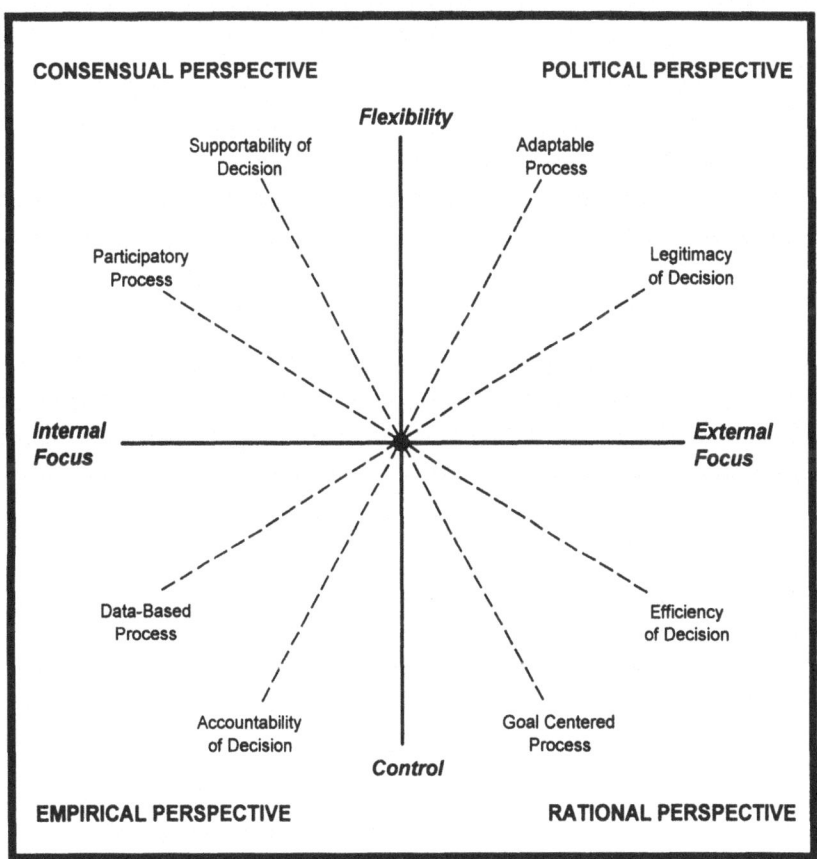

Source: Adapted from Quinn & Rohrbaugh, 1983.

Each criterion is discussed briefly below.

The rational perspective emphasizes goals and objectives. A process is *goal-centered* to the degree it focuses resolutely on the primary problem. In many facility-siting cases, however, problems and goals of the various stakeholders (e.g., generators, environmentalists, federal, state, and local governments) are contradictory and mutually incompatible. To evaluate a siting process, a value-laden position has to be taken regarding the perspective of the analysis. This study analyzes siting processes from the perspective of those responsible for LLRW management. This perspective is quite different from that which would be taken by some of the other stakeholders, for example, by environmental groups opposed to the nuclear power industry.

Goal-centered processes help to reach *efficient decisions*. Assessing the efficiency of decisions is theoretically and practically difficult. For purposes of the present study, we will assume that efficiency is positively related to the degree of progress made in siting a new facility and negatively related to the resources required to achieve this progress.

The empirical perspective emphasizes data and information. A process is *data-based* if decisions are made using sufficient and accurate information that is pertinent to the problem. To assess whether a process is data-based, the quality and quantity of available information, as well as the way it was used in the process, have to be considered.

Data-based processes help to reach *accountable decisions*. A decision is accountable if it is clear, well-documented, and can be readily justified. Processes relying on the comprehensive use of verifiable data and formal models, and those maintaining records of each important step, usually result in accountable decisions.

The consensual perspective emphasizes participation. In a *participatory process*, opinions of all key stakeholders are considered in each phase of the decisions. Steps are taken to assure that these opinions are made known to the responsible decision makers and have considerable influence on the outcome of the process. Forms of participation include direct participation (e.g., public hearings, public inquiries, and referenda), representation (e.g., by elected officials, advisory committees, local monitoring committees, and so forth), or elicitation of opinions (e.g., by polls or surveys). Facilitation may mean providing information, expert support, or financial resources for independent analysis to stakeholders affected by, but not responsible for, the siting. Affected stakeholders may have no impact (merely being kept informed); they may have power to review the work of others and recommend alternative courses of action; or, they may have power to make final decisions.

Participatory processes help to reach *supportable decisions*. Decisions are supportable if they are accepted by the key stakeholders affected by the decisions. If many key stakeholders do not want the facility, a siting decision is not supportable. Defining the concept of key stakeholder is crucial in facility-siting decisions because the population of a local community, a county, a state, or a larger region may have divergent interests and opinions.

The political perspective encourages flexibility and creativity in approaches to a problem. A decision process is *adaptable* if, in response to unexpected events and interventions, it can be easily changed. An adaptable process offers a high degree of flexibility in choosing decision alternatives, criteria, decision rules, and models.

Adaptable processes help to reach *legitimate decisions* that inspire general confidence in their fairness. Decisions are legitimate if they are accepted by a broader public -- along with those directly affected -- even under changing political circumstances.

We use the Competing Values framework to compare and contrast efforts to site LLRW disposal facilities in various states and countries. This framework also organizes our attempt to extract general lessons regarding the factors that influence the success and failure of siting processes.

STUDY METHOD AND STRUCTURE

Case histories of the LLRW disposal facility siting processes in New York, California, Illinois, Nebraska, Texas, Canada, France, the Netherlands, Sweden, and Switzerland were drawn from public documents including legislation, reports generated by government agencies and private companies, summaries of public hearings, letters from interested parties to siting officials, public statements from citizen activist groups, and media reports. The selected states and countries were chosen for more detailed study because preliminary investigations had indicated that they encompassed a wide range of various approaches or were somehow noteworthy or distinctive.

In addition, interviews were conducted with key participants in the LLRW disposal facility siting processes in North American locations -- in New York, California, Illinois, Nebraska, Texas, and Canada. These participants included legislative staff, officials from the relevant government agencies, local officials, generators, leaders of citizen activist and environmental groups, and members of the press. Over 30 interviews were conducted in New York; ten interviews were conducted in California; nine in Illinois; eight in Nebraska; ten in Texas; and six in Canada. Information gathered from the interviews was used to fill gaps where there was not adequate documentation describing the process.

To promote organization and comparability, each case history was divided into seven parallel sections dealing with the major issues that were identified from the documents and interviews:

- Overall Approach to the LLRW Problem;
- Institutional Structure;
- Design of the Site- and Method-Selection Processes;
- Site Selection for the Permanent Disposal Facility;
- Permanent Disposal Method Selection;
- Compensation and Incentives; and
- Interim Waste Management.

After discussion of the case in terms of each of these seven key issues, each case history is concluded by a brief overall analysis from the perspective of the Competing Values framework.

The case histories are grouped into two sections. The first contains the chapters dealing with the five states within the United States. Because the siting process in the various states has been significantly affected by developments at the federal level, the first chapter in this section briefly reviews the history of the LLRW policy at the federal level of government. The second section contains the chapters that describe the siting processes in the five additional countries that we studied.

The concluding chapter attempts to draw some general lessons from the case studies; analyze and compare the processes from the perspective of the competing values approach; and propose recommendations for the design of siting processes.

Chapter 2

FEDERAL HISTORY

A chronological summary of major events related to the federal history of low-level radioactive waste (LLRW) policy is given in Table 1.

Table 1.   A CHRONOLOGICAL SUMMARY OF MAJOR EVENTS RELATED TO THE FEDERAL HISTORY PERTAINING TO LLRW

Date	Event
1946	Ocean dumping of radioactive waste began.
1954	The Atomic Energy Act passed creating the Atomic Energy Commission (AEC).
1959	The Atomic Energy Act was amended creating the "agreement" state program.
1960	The AEC began accepting commercial waste for burial.
1962	Shallow land burial became the preferred disposal method. Commercial LLRW disposal facilities opened at Beatty, NV, and Maxey Flats, KY.
1963-1967	Commercial LLRW disposal facilities opened at West Valley, NY; Richland, WA; and Sheffield, IL
1970	Ocean dumping of radioactive waste stopped.
1971	A commercial LLRW disposal facility opened at Barnwell, SC.
1975-1978	Commercial LLRW disposal facilities at West Valley, Maxey Flats, and Sheffield closed.
1979	The LLRW facilities at Richland and Beatty closed temporarily.
1980	Congress enacted the Low-Level Radioactive Waste Policy Act.
1982	The NRC issued 10 CFR Part 61 regulations.
1985	Congress enacted the Low-Level Radioactive Waste Policy Amendments Act.
1992	U.S. Supreme Court found the take-title provision of the Amendments Act unconstitutional, but left intact all other provisions of the Act. Closure of Beatty and loss of access to Richland sites announced, effective January 1, 1993; projected closure date of Barnwell site to out-of-state LLRW extended through mid-1994.

According to federal legislation,[1] LLRW is radioactive material that is not defined by the Atomic Energy Act of 1954[2] as high-level radioactive waste, transuranic waste,[3] spent nuclear fuel, or by-product material, such as uranium or thorium mill tailings, and that is classified by the Nuclear Regulatory Commission (NRC)[4] as LLRW.   There is substantial heterogeneity in the materials classified as LLRW; designation as LLRW does not necessarily imply low levels of hazard (Gershey et al., 1990; Resnikoff, 1987), although LLRW with high levels of radioactivity tends to be short-lived.

Many processes involving radioactive materials generate radioactive waste. Typical LLRW includes protective clothing, paper and cloth cleaning materials, filters, discarded tools and equipment, solidified liquids, scintillation wastes, animal carcasses, contaminated soil, and activated metals, among others (White & Spath, 1984; Berkovitz, 1987).

The annual national production of commercial LLRW for 1984 to 1990, as measured by the waste received at disposal facilities, is presented in Table 2.   For this period, the annual average volume was approximately 1.85 million cubic feet and the average level of radioactivity was about 500,000 curies.   There has been substantial year-to-year variability in both measures of production, especially the level of radioactivity.   Since passage of the LLRW Policy Act of 1980, the level of waste, in terms of both volume and activity, has been declining.   LLRW accounts for nearly 85 percent of the volume of all radioactive wastes, but less than one percent of the radioactivity (DOE, 1988). In 1990, utilities in the United States generated 56.2 percent of commercial LLRW by volume; industry generated 31.2 percent; and the remainder was generated by non-defense-related federal institutions, academic institutions, and hospitals and medical institutions (DOE, 1991).[5]   In terms of levels of radioactivity, the nuclear power industry was responsible for 79.1 percent of the waste and industry was responsible for 18.7 percent.

Before passage of the LLRW Policy Act of 1980, disposal of LLRW had been subject to a patchwork of federal and state laws and regulations. In the years just after World War II, amounts of LLRW were comparatively small

---

[1]   LLRW Policy Act of 1980 (Pub. L. 96-573) and LLRW Policy Amendments Act of 1985 (Pub. L. 99-240).

[2]   Section 11e.(2), 42 U.S.C. 2014(e)(2).

[3]   Transuranic waste consists of materials contaminated by radionuclides with atomic numbers greater than uranium (e.g., plutonium, curium, etc.).

[4]   Section 61.55, Title 10, Code of Federal Regulations, Part 61.

[5]   The LLRW Policy Act of 1980 and LLRW Policy Amendments Act of 1985 exclude from consideration LLRW generated or owned by the Department of Energy, the U.S. Navy, and defense-related activities. Defense-related wastes account for more than 70 percent of all radioactive wastes generated in the United States (DOE, 1988).

Table 2. VOLUME OF WASTE AND ACTIVITY RECEIVED BY
DISPOSAL FACILITIES

	Volume (Cubic feet)	Radioactivity (Curies)
1985	2,680,565	748,874
1986	1,805,503	233,740
1987	1,841,637	269,550
1988	1,427,850	259,661
1989	1,625,862	866,868
1990	1,142,810	547,902

Source: DOE, 1991

by today's standards. The Atomic Energy Act of 1954 assigned responsibility for managing both high- and low-level radioactive waste to the Atomic Energy Commission (AEC). In the early years, the focus was primarily on managing waste produced by defense-related activities (O'Connor & Shaw, 1988). The AEC disposed of LLRW, most frequently, by ocean dumping (DOE, 1984/b; White & Spath, 1984; Resnikoff, 1987).

From the time of the Manhattan Project, LLRW was also disposed of at five project sites: Hanford, Washington; Los Alamos, New Mexico; Savannah River, Georgia; Idaho Falls, Idaho; and Oak Ridge, Tennessee (O'Connor & Shaw, 1988). At these sites, the disposal method was shallow land burial, in which packaged wastes, contained in cement casks, steel drums, or wood or cardboard bins, are buried in shallow trenches, generally within 30 meters of the surface.

By the early 1960s, shallow land burial had displaced ocean dumping as the preferred disposal method. Public concerns about ocean dumping led the AEC to stop issuing new ocean disposal licenses in 1960. In 1960, AEC began to accept commercial waste and buried it along with military and other federal wastes at Oak Ridge and Idaho Falls. By 1962, 95 percent of LLRW was being disposed of by shallow land burial (White & Spath, 1984).

The 1959 amendments to the Atomic Energy Act permitted states to apply to the AEC for the authority to administer their own radiation protection programs. States authorized by AEC were called "agreement states," and they could regulate land disposal, among other activities. In 1962, the AEC began accepting license applications for private companies to operate LLRW sites. These regional sites were to be located on either federal or state lands, but could be operated by commercial firms under license from AEC or an agreement state.

The first two commercial LLRW disposal sites opened in 1962 in Beatty, Nevada, and Maxey Flats, Kentucky. During the next decade, four other disposal sites opened: West Valley, New York, in 1963; Richland, Washington, in 1965; Sheffield, Illinois, in 1967; and Barnwell, South Carolina, in 1971. By 1978, however, the West Valley, Maxey Flats, and Sheffield sites were closed. Environmental problems, related primarily to water infiltration, flooding of the trenches, and the migration of water and radioactive materials from the trenches, were observed at each of the three closed sites.

After the other three sites closed, all the nation's commercial LLRW was disposed of in the remaining three sites at Barnwell, Beatty, and Richland. The governors of South Carolina, Nevada, and Washington objected to bearing the LLRW disposal burden for the entire nation, and they urged development of a system of regional disposal sites.

In 1979, both the Richland and Beatty sites were temporarily closed by their respective states' governors in response to operational problems. In South Carolina, a policy was established to limit by half the amount of waste the Barnwell site would accept. In 1980, the voters of Washington approved an initiative that banned out-of-state waste from the Richland site, although this ban was later ruled unconstitutional (Berkovitz, 1987; Gershey et al., 1990; Resnikoff, 1987; White & Spath, 1984).

In 1980, the National Governors' Association, the National Conference of State Legislatures, and the State Planning Council on Radioactive Waste Management, which had been formed by President Carter, all recommended giving responsibility for managing LLRW to the states. At the end of its 1980 session, Congress enacted the LLRW Policy Act, which made each state responsible for disposing of LLRW generated within its own boundaries. States were encouraged to enter into compacts to provide for establishing and operating regional disposal facilities, although these compacts had to be ratified by Congress. After January 1, 1986, compacts could exclude LLRW generated outside the region. The law did not prohibit states from "going it alone," but, based on the commerce clause of the Constitution (Berkovitz, 1987; Gershey et al., 1990), some observers questioned whether such states would be able to exclude waste from other states or regions.

In December 1982, NRC issued Title 10, Code of Federal Regulations, Part 61, "Licensing Requirements for the Land Disposal of Radioactive Waste." Along with related amendments to other NRC regulations, Part 61 represented the first comprehensive set of criteria for regulating LLRW disposal in the country. The regulations specify the means to process, package, and dispose of LLRW. All agreement states must abide by these regulations, although they may impose additional restrictions.

These regulations subdivide LLRW into three classes, A, B, and C, depending on its radionuclide source, half-life, concentration, and energy level. In 1987, Class A accounted for 97.4 percent of the volume and 9.6 percent of the radioactivity of LLRW shipped to commercial disposal sites; Class B accounted for 2.1 percent of the volume and 24.9 percent of the radioactivity; and Class C accounted for .5 percent of the volume and 65.5 percent of the radioactivity (DOE, 1988).

In general, wastes are regulated more stringently depending on their level of hazard, which is determined by concentrations of radioactivity, the half-life of the nuclides, and mobility (the solubility or volatility of the radionuclide and the waste form). The maximum concentration limits for LLRW from 10 CFR Part 61 are given in Table 3.

The regulations governing the treatment and disposal of LLRW are progressively stricter for each class. However, all classes of LLRW must meet the same minimum requirements before disposal. For example, LLRW cannot be packaged in cardboard containers, liquids must be solidified or packed with sufficient material to absorb twice the volume of liquid waste, and solid wastes may contain no more than one percent free-standing liquids.

Class A wastes have the lowest concentrations of overall radioactivity, with only extremely small amounts of long-lived radionuclides. Class B has higher concentrations of radioactivity and must meet stricter standards for compression, leaching, and degradation. Class C waste contains the highest concentrations of radioactivity and must meet the same stability standards as Class B waste in addition to being disposed of in a manner that protects against accidental exposure by inadvertent intruders for 500 years. LLRW exceeding the concentrations of radioactivity permitted for Class C waste is classified as "Greater than Class C" (GTCC) and is prohibited from commercial LLRW disposal sites.

10 CFR Part 61 contains procedures, performance objectives, and technical criteria for licensing facilities for land disposal of LLRW. The regulations specify the general requirement that land disposal facilities must be sited, designed, operated, closed, and controlled after closure so that reasonable assurance exists that exposures to humans are within certain specified limits. The regulations identify four specific performance objectives: to protect the general population from releases of radioactivity; to protect individuals who inadvertently intrude after facility closure; to protect individuals during operations; and to stabilize the disposal site after closure.

Table 3. MAXIMUM CONCENTRATION LIMITS FOR LOW-LEVEL RADIOACTIVE WASTE FROM 10 CFR PART 61

Radionuclide Nuclides with Half-lives	Half-Life (years)	Maximum Concentration Limits $(Ci/m^3)^a$		
		Class A	*Class B*	*Class C*
<5 years[b]	<5.0	700.00	NL[c]	---[d]
^{60}Co	5.3	700.000	NL	---
^{3}H	12.3	40.000	NL	---
^{90}Sr	28.0	0.040	150.0	7000.00
^{137}Cs	30.0	1.000	44.0	4600.00
^{63}Ni	92.0	3.500	70.0	700.00
^{63}Ni in activated metal	92.0	35.000	700.0	7000.00
^{14}C	5,730.0	0.800	---	8.00
^{14}C in activated metal	5,730.0	8.000	---	80.00
^{94}Nb in activated metal	20,000.0	0.020	---	0.20
^{59}Ni in activated metal	80,000.0	22.000	---	220.00
^{99}Tc	212,000.0	0.300	---	3.00
^{129}I	17,000,000.0	0.008	---	0.08
alpha-emitting transuranic nuclides with half-lives <5 years	<5.0	10.000 nCi/g	---	100.00 nCi/g
^{242}Cm	0.45	2,000.000 nCi/g	---	20,000.00 nCi/g
^{241}Pu	13.2	350.000 nCi/g	---	3,500.00 nCi/g

^{a} Ci = 3.7 x 10^{10} Bq
^{b} Including but not limited to: ^{32}P, ^{35}S, ^{51}Cr, ^{54}Mn, ^{55}Fe, ^{58}Co, ^{59}Fe, ^{65}Zn, ^{67}Ga, ^{125}I, ^{131}I, ^{134}Cs, ^{144}Ce, and ^{192}Ir.
^{c} No upper limit on concentration.
^{d} Class not defined for these nuclides.

Source: Gershey et al., 1990

Suitability requirements for LLRW disposal facilities -- the minimum characteristics a disposal site must have to be acceptable for use as a near-surface facility -- are also specified. All states must satisfy these requirements, although they are free to impose additional suitability requirements, and many have done so.[6]

Within a few years after the passage of the 1980 Act, the three states with existing sites quickly established compacts with other states. The other states, including many of those that were among the largest generators of LLRW, moved more slowly.

By 1984, although several compacts had been formed by states, none had been ratified by Congress. The states without sites had little incentive to support ratification of the proposed compacts. Doing so would have led to the possibility that their own waste would have been excluded by the very compacts that they had approved. Even for the states that had agreed to enter compacts, it became apparent that they would be unable to establish operational sites before January 1, 1986, because of technical and political constraints.

Representatives from the states with operating disposal facilities indicated that their states would agree to accept LLRW from outside their compact regions for some limited period after the January 1, 1986, deadline if Congress would ratify their compacts. They indicated that their acceptance of such waste would be conditioned on clear progress by other states in the development of new disposal facilities; an absolute cut-off date for access to sites by out-of-state generators; immediate reductions in the amount of waste to be disposed of; and Congressional ratification of their compacts (Brown, 1986).

After extended debate, negotiation, and hearings, the LLRW Policy Amendments Act of 1985 was passed. The key provisions of the Act were that the three currently operating sites agreed to remain open for an additional seven years; the states and regions without identified disposal facility sites agreed to accept a number of eligibility conditions to retain access; the three states containing disposal facility sites would not have to accept waste in excess of specified ceilings; generators in states and regions without their own disposal facilities would be subject to paying escalating surcharges on their waste; and states and regions without disposal facilities would have to meet specific milestones in 1986, 1988, 1990, and 1992 or

---

[6] The minimum requirements include that the site must be able to be characterized, modeled, analyzed, and monitored; should not fall within areas that will probably be affected by future population growth and development; should not contain exploitable natural resources, such as ore or petroleum reserves; should be well drained and free of frequent flooding or ponding; should have a minimum upstream drainage to minimize the possibility of erosion or inundation from runoff; may not be within a zone where groundwater intrusion is possible; should not be in areas of high seismic, volcanic, erosion, or mass wasting activity; and should not be located near other facilities or activities that may adversely affect the site or that have emissions that make monitoring the environmental impacts of the site difficult. (Summary adapted from Gershey et al., 1990).

would have to pay penalty surcharges and might be denied access. (This summary of the law's key provisions is adapted from Brown, 1986.).

The Act required states to take title to and possession of waste, and to assume liability for any damages to a generator caused by failure to do so, if the state were not able to provide disposal capacity for the generators of waste by January 1, 1996 (Berkovitz, 1987). The Act charged NRC to develop standards and procedures for exempting specific waste streams so low as to be "below-regulatory-concern," but final regulations have yet to be issued. A gap in federal regulatory policy also exists for mixed waste, which consists of LLRW waste that is also hazardous. Such waste is estimated to account for two to five percent of total LLRW (Gershey et al., 1990).

In February 1990, the State of New York, Cortland County, and Allegany County filed a lawsuit against the federal government, challenging the constitutionality of the LLRW Policy Act Amendments, especially the take-title provision. The challenge was based primarily on the 10th Amendment, which guarantees that "powers not delegated to the United States by the Constitution, nor prohibited it by the States, are reserved to the States respectively, or to the people."

In December 1990, the Federal District Court issued a ruling upholding the constitutionality of the federal Amendments Act of 1985. In June 1991, the United States Court of Appeals for the Second Circuit upheld the opinion of the district court. In June 1992, however, the Supreme Court (*NY v. United States et al.*, 1992) found the take-title provision of the Amendments Act unconstitutional, but left intact all other provisions of the Act.

In anticipation of the eventual Supreme Court decision, progress towards facility siting slowed everywhere and, as of the end of 1992, has remained stalled or slowed partially because of uncertainty about possible future Congressional actions. The Beatty disposal facility closed as of January 1, 1993. Also as of that date, the Richland facility will remain open only for LLRW from the Northwest and Rocky Mountain Compacts. In June 1992, the South Carolina Legislature extended the projected closure date for the Barnwell facility through 1995. In Autumn 1992, the Southeast Compact Commission indicated that it would continue to accept out-of-compact waste through June 30, 1994, and announced sharply increased access fees.

Chapter 3

NEW YORK

BACKGROUND

New York State is a relatively large producer of LLRW. In 1990, New York State ranked fifth nationally in the volume of waste shipped to LLRW disposal facilities and seventh in the amount of radioactivity shipped. Seventy-two percent by volume of this waste was from utilities, 28 percent from institutional and industrial sources, and the remaining fraction was from government. Table 4 gives the total LLRW from New York State disposed of at commercial disposal sites for the period 1979 to 1990. The large majority of LLRW is shipped to the Barnwell, South Carolina, facility; in 1989, a typical year, over 70 percent of the waste by volume and over 98 percent of the waste by activity was shipped there.

From 1963 to 1975, New York State contained one of the nation's six commercial LLRW facilities. It was operated by Nuclear Fuel Services, Inc., (NFS) at the Western New York Nuclear Service Center near West Valley. During this period, the West Valley site accepted 2.3 million cubic feet of LLRW for disposal. Waste was placed in long trenches which were covered with excavated earth. In March 1975, the water seeped through the covers of two trenches leading to discontinuation of waste disposal at the West Valley site. NFS took remedial action (Anderson, 1987; Matuszek, 1988), and responsibility for the management of the site was transferred to the New York State Energy Research and Development Authority (Energy Authority) in 1983.

OVERALL APPROACH

A chronological summary of major events related to the overall approach to the LLRW problem in New York State is given in Table 5.

In 1981, shortly after the federal LLRW Policy Act of 1980 was passed, the Coalition of Northeastern Governors (CONEG) formed a Compact Task

Table 4. TOTAL LLRW GENERATED IN NEW YORK AND DISPOSED
OF AT COMMERCIAL DISPOSAL SITES BY YEAR

	Volume (cubic meters)	Radioactivity (curies)	Volume Percentage by Source		
			Institutional and Industrial	Power Reactor	Government and Military
1979	9,572	78,961	61%	32%	7%
1980	7,283	63,406	51%	46%	3%
1981	7,021	71,878	47%	53%	0%
1982	6,677	29,164	23%	77%	0%
1983	5,622	70,192	42%	58%	0%
1984	4,156	30,008	39%	61%	0%
1985	4,563	17,131	42%	56%	2%
1986	3,017	9,410	80%	20%	0%
1987	1,993	5,060	32%	65%	3%
1988	2,069	20,867	25%	75%	0%
1989	2,744	101,454	25%	67%	8%
1990	2,019	12,701	28%	72%	0%
1991	3,124	103,140	26%	74%	0%

Source: For all years except 1991; DOE, 1980 - 1991. For 1991;
NYSERDA, 1992/a.

Force, including representatives of all the New England states, New York,
New Jersey, Pennsylvania, Delaware, and Maryland, to begin discussions
[7]about forming a northeast compact agreement.

By Spring 1983, the Compact Task Force had developed a proposed
compact specifying operational responsibilities, provisions for liability,
administrative and legal arrangements, and the criteria for selecting a host
state. In August 1983, legislation[7] was passed requiring the New York State
Energy Office (NYSEO) to study the potential advantages and disadvantages
of joining a compact with other states versus siting a facility that would
handle only LLRW generated within the State. To provide assistance to the
study, the law established an Advisory Committee, including legislators,
members of State agencies, local authorities, environmental groups, and
generators.

[7]   The Low-Level Radioactive Waste Disposal -- Study Law (NYS Ch. 978, 1983) amended
      section 18 of the Energy Law.

Table 5. CHRONOLOGICAL SUMMARY OF KEY EVENTS RELATED
TO THE OVERALL APPROACH TO THE LLRW PROBLEM

Date	Event
1981	The Coalition of Northeastern Governors (CONEG) formed a Compact Task Force to discuss the formation of a Northeast Compact. Generators formed the New York State Low-Level Waste Group.
1983	The NYS Legislature enacted the Low-Level Radioactive Waste Disposal-Study Act requiring the New York State Energy Office (NYSEO) to make recommendations on the proposed Compact.
1984	The NYSEO report recommended NYS reject the Northeast Compact.
1985	Governor's Program Bill #54, proposing legislation to implement the federal mandate, was submitted to the legislature.
1986	The NYS Legislature enacted the LLRW Management Act.
1989	Communities and activist groups mounted strong opposition to siting process.
1990	The Legislature enacted the LLRW Management Act Amendments and passed legislation requiring that title to LLRW remain with generators. NYS and Corland and Allegany counties filed a lawsuit against the federal government to overturn portions of 1985 Amendments Act.
1992	The U.S. Supreme Court overturned the take-title provision of the 1985 Amendments Act.

In April 1984, after input from the Advisory Committee and a series of public hearings, NYSEO issued a report on LLRW management (NYSEO, 1984/b) that analyzed the options in terms of health and safety impacts; environmental impacts; economic impacts; and policy and social impacts. According to the study, the provisions of the proposed CONEG compact strongly favored selecting New York State to host a site, without guaranteeing rotation of host state responsibilities among compact members. The analysis recommended that New York State should attempt to negotiate a compact with one or more northeast states because the LLRW Policy Act of 1980 explicitly enabled only those States that were members of compacts to exclude LLRW from other states. The report concluded, however, that if the disposal facility were State-owned and operated, the State could probably constitutionally limit disposal to waste generated within the State. The report recommended enacting legislation to establish a process to site a permanent disposal facility in New York.

Governor's Program Bill #54 was submitted to the Legislature in March 1985. It proposed that the Energy Authority establish a temporary above-ground, engineered, monitored storage facility for non-utility LLRW at the Western New York Nuclear Service Center in West Valley, which would be closed after completion of new permanent disposal facilities. The New York State Department of Environmental Conservation (NYSDEC) was to conduct a detailed study that would identify up to three sites for permanent LLRW disposal facilities and appropriate disposal methods for each site. Following final selection of a site(s) and disposal method(s), the Energy Authority would be required to construct and operate the permanent disposal facilities. Both the temporary and permanent facilities would be financed through fees from LLRW generators. An advisory committee would be created to work with NYSDEC and the New York State Department of Health (NYSDOH) to plan and carry out a Statewide public information program.

The Governor's Program Bill was amended in the Assembly. The West Valley site was specified as the location of the temporary storage facility but prohibited from being a site for a permanent disposal facility. The Assembly bill also specified creation of a five-member commission for siting low-level radioactive waste disposal facilities (NYS Assembly Bill 7227--B, 1985). Eventually, the issue was tabled until the next legislative session.

The federal LLRW Policy Amendments Act was passed in December 1985, extending the deadlines of the 1980 legislation and making it clear that the State would probably not lose access to disposal facilities outside the State for several years. The new federal law influenced legislative deliberations during the 1986 session.

In March 1986, Governor's Program Bill #167 on LLRW was introduced, and legislative hearings were held. The bill proposed a comprehensive LLRW management program and a process for siting, constructing, and operating a disposal facility exclusively for LLRW produced by generators in New York State. It called for establishing LLRW management facilities by January 1, 1993. It made no mention of a compact arrangement, although it required that efforts be made before constructing a facility to determine whether disposal capacity for State waste was available elsewhere. At the same time, it precluded accepting out-of-State waste in a State facility.

The bill prescribed that all expenses related to the selection, development, licensing, constructing, operation and maintenance of the disposal facility, including financial aid to hosting communities and compensation for the damages of potential accidents, should be administered by the State and covered by fees collected from generators. The bill strengthened the take-title provision from the federal law, dictating that the State take title to the waste accepted for permanent disposal.

During the hearings, the proposed bill was widely supported because, without enacting the bill, New York State would not be able to comply with federal law and might be excluded from existing facilities. Some environmental groups, however, urged the Legislature to exempt Class C wastes from LLRW facilities, arguing that NRC's 100-year institutional control period does not protect the public from long-lived radioactive wastes (*Senate Journal*, July 1986, p. 30).[8] Environmental groups also criticized

---

[8]   The NRC regulations, in fact, prohibit reliance on uncertain institutional controls beyond 100 years, requiring other means of protections, such as physical barriers, depth of cover, and so on, after this time.

the take-title provision, arguing that no other industry producing hazardous waste gets this type of support.

The New York State LLRW Management Act was passed into law and became effective in July 1986. A bill to regulate the transportation of Low-Level Radioactive Waste in New York State was passed at the same time (NYS Ch. 508, 1986).

The LLRW Management Act established a Siting Commission to oversee the site and method selection processes. After several years of work, the Commission identified ten candidate areas for a facility site in December 1988. Local officials and citizen groups then protested the siting process. In September 1989, the Siting Commission further narrowed the siting to five potential sites, three in Allegany County and two in Cortland County. In November 1989, after a meeting with officials and concerned citizens from Allegany and Cortland counties, where potential sites were located, the Governor promised to pursue possibilities for an interstate compact and other means of obtaining access to out-of-State facilities, and said he would challenge the Federal law on the take-title issue (Cuomo, 1989/a).

The Governor sought to have Congress revise the federal law, contacting members of the New York State delegation and Chairmen of the appropriate committees and exploring possibilities for using the National Governor's Association as a vehicle to stimulate interest. Congress has been unwilling, however, to revisit the 1980 and 1985 LLRW Policy Act and Amendments because of the general assumption that, if Congress indicated a willingness to revise the law, existing siting processes would come to a halt (Berkowitz, 1990).

In November 1989, the Governor wrote to all states in the process of siting a facility to explore their interest in accepting waste from New York State after January 1993. The response was generally negative, although responses from California (Duekmajian, 1989) and Texas (Clements, 1990) appeared to leave the possibility open.

To remain in compliance with current federal law, in December 1989, the Governor sent to NRC, DOE, and the states of South Carolina, Nevada, and Washington a letter of certification accompanied with the New York State plan for managing LLRW after 1992. In January 1990, the Governor was advised by the three states that New York State was found to be in compliance with the federal law.

In February 1990, several local citizens' groups presented legislators with a series of proposals regarding New York State's radioactive waste problem,[9] some of which were later included in the amended law. Also in

---

[9] These included eliminating the take-title provision of federal and State law; investigating and auditing the Siting Commission; mandating zero-release of radionuclides from waste; appropriating funds to counties containing potential sites; restricting the use and acquisition of agricultural lands; further studying earthquake faults, groundwater, and aquifers;

February 1990, the State of New York, Cortland County, and Allegany County filed a lawsuit against the federal government challenging the constitutionality of the LLRW Policy Amendments Act, especially the take-title provision (United States District Court, 1990).

In March 1990, the Governor proposed legislation that eliminated the take-title provision from the State law. In July 1990, legislation was passed requiring that title to waste remain with the generators and making other important changes in the institutional arrangements and the siting procedure, discussed below.

The legislated changes in the siting procedure led to suspension of the site-selection process until a preferred disposal method was chosen. In light of this development, the states of Nevada, South Carolina, and Washington sent a letter of concern to the Governor in October 1990, asking for additional information to demonstrate that New York State would meet the federal mandate to manage the State's LLRW after January 1993. At the direction of the Governor's office, a detailed report was prepared by the Energy Authority and sent to the states in December 1990 accompanied by a letter from the Governor assuring them of New York State's good faith efforts to meet the federal mandate. Officials of the three states found New York State to be in compliance with the federal law and assured the Governor that they would allow New York State generators continued access to disposal facilities within their boundaries.

In response to the New York State lawsuit, in December 1990, the Federal District Court issued a ruling upholding the constitutionality of the federal Amendments Act of 1985. In June 1991, the State of New York, Cortland County, and Allegany County appealed to the United States Court of Appeals for the Second Circuit, which upheld the opinion of the district court. New York State filed a petition for the U.S. Supreme Court to hear the case, September 1991. In June 1992, the Supreme Court (*NY v. United States et al.*, 1992) found the take-title provision of the Amendments Act unconstitutional, but left intact all other provisions of the Act.

INSTITUTIONAL STRUCTURE

A chronological summary of major events related to the institutional structure in New York State is given in Table 6.

---

increasing the degree of local municipal control in the siting process; and banning incineration, interim storage, and the disposal of LLRW in New York State (Cortland Citizens Against Radioactive Dumping et al., 1990/a).

Table 6. CHRONOLOGICAL SUMMARY OF KEY EVENTS RELATED
TO THE INSTITUTIONAL STRUCTURE

Date	Event
1962	NY was granted "agreement" state status.
1983	The Energy Authority took over management of the West Valley site.
1984	The NYSEO report recommended that NYSDEC conduct the siting process and the Energy Authority operate the disposal facility
1986	The Siting Commission was created to conduct the siting process with the help of an Advisory Committee. NYSDOH instituted a public information program.
1987	Siting Commission members were appointed.
1990	Siting Commission membership was expanded and the Advisory Committee was changed to the Citizen's Advisory Committee.
1991	New members of the Citizen's Advisory Committee were appointed. NYSDOH developed a draft scope of work to establish the Scientific and Technical Review Panel.

As an agreement state, since 1962 New York State has possessed the authority to license all uses of radioactive material except for defense-related activities and the licensing of electric utilities.   This authority has been divided among the NYSDOH and the New York City Health Department (NYCHD), which regulate the use of radioactive material by medical, governmental, and academic institutions; the New York State Department of Labor (NYSDOL), which regulates the use and handling of radioactive material by commercial and industrial entities; and the NYSDEC, which regulates the siting, design, construction, operation, closure, monitoring, and safety planning of LLRW disposal facilities, as well as any radioactive discharges to the environment.   NYSEO coordinates the Agreement State program and acts as a liaison with the NRC.

At the request of the State Legislature, NYSEO investigated the question of the appropriate institutional structure for siting a permanent disposal facility in New York State.   The 1984 NYSEO report recommended that NYSDEC conduct the siting process, with an advisory committee consisting of representatives of state agencies, the public, and waste generators.   The Energy Authority would be responsible for preparing and submitting the required applications for licensing the facility, as well as for the construction, operation, and post-closure activities associated with the facility (NYSEO, 1984/b).

Several alternative approaches to the siting process were widely discussed.   One option was to assign responsibility for conducting the siting process to NYSDEC, as had been recommended by NYSEO, but NYSDEC took the position that its role was to regulate a facility, not promote it.

Another option was to assign responsibility to the Energy Authority because they would own and operate the facility, but the Energy Authority objected that this would leave them open to charges that their interests related to managing the facility might unduly influence decisions on site and disposal method selection. A third option was to use a private company to site and operate the LLRW facility, which was opposed by many people because previous experience with private operation at West Valley had been problematic and because there was concern that if New York State did not join a compact, this option might make it impossible to exclude outside waste. A fourth option was for an independent state agency to conduct the siting process.

The NYSLLRW Management Act, enacted in July 1986, assigned responsibilities for siting, constructing, and operating a disposal facility. The bill mandated the establishment of an independent Siting Commission for conducting site and method selection. The Siting Commission was required to prepare a draft environmental impact statement and apply to NYSDEC for certification of site(s) and method(s). Members of the Commission would be appointed by the Governor and would include a geologist, a medical doctor, a health physicist, a professional engineer, and a private citizen who would act as chairperson.

The Energy Authority was directed to design and operate the disposal facility; prepare a draft environmental impact statement on facility design, construction, operation, and closure; apply to NYSDEC for permits to build, operate, and close the facility; and apply to NYSDOL for license to handle radioactive materials. The Energy Authority was also required to collect information and provide annual reports to the Governor and Legislature on the types and quantities of LLRW generated within the State and to collect annual assessments from operating nuclear power plants and manage other financial matters.

NYSDEC was required to specify criteria for disposal facility siting; prepare a final environmental impact statement(s) on site(s) and disposal method(s); and certify compliance with regulations. NYSDEC was also required to prepare a final environmental impact statement on facility design, construction, operation, and closure, and to approve or deny permits for construction and operation. NYSDEC was to promulgate regulations on financial assurance requirements for LLRW disposal facility operators and to make recommendations about the nature and form of State assistance to local communities.

NYSDOH was required to conduct a statewide public information program on the health and safety implications of LLRW management. The law also required the creation of the LLRW Advisory Committee, which was to include the State geologist; the commissioners or designees of the Departments of Health, Labor, Transportation, State, and the Energy Office;

two representatives of non-profit environmental organizations; two health physicists or medical doctors; two representatives of LLRW generators; and one private citizen. Three private citizens would be appointed to the Advisory Committee from the counties where proposed sites were identified. The members and the chairman of the Committee were to be appointed by the Governor. The purpose of the Advisory Committee was to advise the Siting Commission on site and method selection, NYSDEC on its LLRW regulatory program, and NYSDOH on its LLRW public information program.

In May 1987, members of the Siting Commission were appointed. The chairman of the Advisory Committee, a health physicist, was appointed in March 1988 (NYSLLRWAC, 1988/b). As originally conceived, the Advisory Committee was to continue to exist until issuance of NYSDEC's final environmental impact statement accompanying the specific low-level radioactive waste site(s) and method certification (NYSLLRWAC, 1988/a).

After the Siting Commission identified candidate areas for siting a disposal facility in December 1988, and potential sites in September 1989, the level of public interest and involvement increased sharply. Local officials from Allegany and Cortland Counties, which contained the potential sites, and citizen groups such as Concerned Citizens of Allegany County (CCAC) and Cortland Against Radioactive Dumping (CARD) protested the siting process, demanded the dismissal of the Siting Commission, and argued that local municipalities should be given the authority to license, permit, and approve or prohibit construction or operation of a LLRW facility within their boundaries (Cortland Citizens Against Radioactive Dumping et al., 1990/a).

After meeting with representatives of Cortland and Allegany counties in May 1990, the Governor submitted a Program Bill that proposed modifying the site- and method-selection process by amending the NYSLLRW Management Act of 1986. The bill's stated objectives were to address citizens' concerns with ample opportunity for public participation at key decision points; provide independent reviews of the process and decisions made to that date and in the future; and to consider thoroughly all reasonable disposal method alternatives and siting options (Governor's Program Bill #317, 1990).

In July 1990, the bill was enacted as the Low-Level Radioactive Waste Management Amendments Act (NYS Ch. 913, 1990). The membership of the Siting Commission was increased from five to seven members by adding a social scientist and an environmentalist to the existing members. NYSDOH was directed to contract with one or more independent panels of technical and scientific experts to review and evaluate the Siting Commission's work to date, as well as any future work. The Siting Commission was directed to make available to the public all computerized data in a format accessible for general use.

The Advisory Committee was reconstituted, becoming the Citizens' Advisory Committee (CAC). It was made autonomous from the Siting Commission; more citizen membership was added; State agency seats were eliminated; and an executive director was to be hired. NYSDOH was directed to provide administrative support to the CAC. The CAC was directed to facilitate public review and comment on the disposal method-selection process, the site-characterization and selection processes, the process of combining site and preferred disposal method, the associated draft environmental impact statements, and the manner in which the Siting Commission makes public information available.

A process was established for selecting a preferred disposal method and establishing method-specific site-selection criteria before resuming site-selection activities. The Siting Commission was required to evaluate all reasonable alternative disposal methods, including deep-vertical-shaft mined disposal and above-ground monitored retrievable disposal, and to issue a generic, non-site-specific environmental impact statement to NYSDEC for certification. The CAC and the independent scientific and technical panel was to be given an opportunity to review and comment on the preferred disposal method-selection process.

The Siting Commission was directed to issue a report concerning all lands previously excluded from consideration and to make available reports and other documents which explain or support its prior decisions relating to site selection and future activities concerning disposal method and site selection. The CAC and the independent scientific and technical review panel was to be given the opportunity to review and comment on these documents, and the Commission was directed to respond to comments received (NYSERDA, 1990/c).

Some citizen groups opposed the amendments because they did not remove the existing potential sites from consideration or address the generation of the waste, but one local government endorsed the amendments because they were perceived to mandate investigation of waste management alternatives, such as on-site storage, and encourage greater accountability and justification in the decision-making process (Cortland County LLRW Office, 1990).

The 1990 Amendments to the State Act led the Siting Commission to develop plans to make its activity more open and visible to the general public (NYSLLRWSC, 1991/a). New members of the Committee were not appointed by the Governor's office until Summer 1991. The CAC had its first meeting in August 1991 and announced that it would first review the siting process in terms of its openness, responsiveness and fairness, and, then, monitor future activities of the Commission (NYSLLRWSC, 1991/b). During 1992, the Siting Commission and its contractors solicited views of the

CAC on various issues, including public participation in method selection, site selection, facility monitoring and oversight (NYSLLRWSC, 1992/a-f).

In August 1991, NYSDOH developed a draft scope of work proposing an independent Scientific and Technical Review Panel to review the Siting Commission reports. NYSDOH recommended contracting with the National Academy of Sciences (NAS), on the grounds that NAS has a rigorous report review process and that selection of experts by NAS would ensure that reports meet scientific standards and are impartial (NYSDOH, 1991/b). NYSDOH solicited comments and input on the draft scope of work from the CAC, state agencies, and the general public. In June 1992, NYSDOH issued a revised scope of work (NYSDOH, 1992/a) and approached NAS to determine the possibility of a contract. As of December 1992, the contracting process was in progress.

The LLRW Management Act required NYSDEC to promulgate financial assurance regulations specifying financial requirements for operators of LLRW disposal facilities (in the present case, the Energy Authority) that would provide for the remediation of failures during operation and after facility closure, and for pre-closure and post-closure facility monitoring and maintenance. In September 1990, NYSDEC published proposed regulations that provide coverage for the costs of closure, post-closure monitoring, maintenance during the institutional control period, remediation of failures, and third-party compensation (NYSDEC, 1990).

A public hearing on the draft regulations was held in December 1990, and oral and written comments were received from the operator, generators, environmental groups, local governments, and the general public (NYSDEC, 1991/a). Final regulations and the final generic environmental impact statement were published in December 1991 (NYSDEC, 1991/b). In December 1991, NYSDEC issued draft regulations on design, construction, operation, closure, post-closure, and institutional control of LLRW disposal facilities (NYSDEC, 1991/c). Four public hearings were held across the State, and written comments were received from individuals and organizations. Final regulations were issued in August 1992 (NYSDEC, 1992).

DESIGN OF THE SITE- AND METHOD-SELECTION PROCESSES

A chronological summary of major events related to the design of the site- and method-selection processes in New York State is given in Table 7.

The 1984 NYSEO study made recommendations about the design of the site- and method-selection processes, based primarily on the NRC 10 CFR

Table 7. CHRONOLOGICAL SUMMARY OF KEY EVENTS RELATED
TO THE DESIGN OF THE SITE- AND METHOD-SELECTION
PROCESSES

Date	Event
1984	The NYSEO report recommended that site selection should be followed by method selection.
1987	NYSDEC issued site and method selection regulations.
1988	The Siting Commission selected Roy F. Weston, Inc., as their contractor. The Siting Commission published the Site- and Method-Selection Plans, the Generic Scope of Draft Environmental Impact Statement, and the Draft Statewide Exclusionary Screening Report.
1990	KPMG Peat Marwick was hired to audit Roy F. Weston, Inc. The Siting Commission was directed to select a disposal method before continuing the site-selection process.
1991	The Siting Commission terminated its contract with Weston, Inc. The Siting Commission selected technical support contractors.

Part 61 regulations, which state that they rely more heavily on the geology of
the site for safety than on the type of construction because the effectiveness
of design features and technology are expected to decrease over time (U.S.
NRC, 1982). The 1984 NYSEO report recommended a two-phased siting
study in which site selection would be followed by the selection of the most
appropriate disposal technology.

The NYSLLRW Management Act of 1986 (NYS Ch. 673, 1986) did not
specify precisely the stages of the site- and method-selection processes, or
their sequence. It encouraged the selection of a single site, but multiple sites
were permissible, as were multiple disposal methods, if necessary and
appropriate. The law directed the Siting Commission to prepare a draft
environmental impact statement on method and site selection to accompany
its application for certification submitted to NYSDEC.

NYSDEC was directed to promulgate regulations on the procedures for
method and site selection, application, and certification. NYSDEC was also
required to hold public hearings and to make a final decision regarding the
acceptability of the application for certification and the final environmental
impact statement.

During development of regulations specifying procedures for method and
site selection, application, and certification, NYSDEC prepared a Draft
Environmental Impact Statement that analyzed regulatory alternatives and
reviewed their potential environmental impacts (NYSDEC, 1987/a).
NYSDEC issued draft regulations in July 1987 and initiated a series of public
hearings. NYSDEC also received a number of written responses from

federal and State agencies and institutions; city, county, and regional agencies or commissions; utilities and corporations; independent organizations; and private individuals. In December 1987, NYSDEC issued 6 NYCRR Part 382: Regulations for LLRW Disposal Facilities, which "establish minimum criteria needed to ensure long-term isolation of LLRW and control exposure to the public and the environment to radiation from the disposal of LLRW" (NYSDEC, 1987/b, Executive Summary, p. 2) and prescribe procedures for the main stages of the process.

According to Part 382 regulations, the Siting Commission must choose one or two proposed sites and, for each site, one or two disposal methods. The Commission must perform site characterization studies, including on-site investigations and monitoring, to demonstrate that the proposed site(s) meets the regulatory requirements and must submit plans for the characterization studies to NYSDEC for prior review. The Commission must consult with NYSDEC regarding the form and content of the application for certification; must submit for NYSDEC's approval its proposed scope before preparing the draft environmental impact statement (DEIS); and must hold a public scoping meeting on the scope of the DEIS. The Commission has to submit copies of the application to NYSDEC, the Energy Authority, the Legislature, and the Governor. NYSDEC has to review the application, hold public hearings, and decide on certification. Proposed sites and methods must be in conformance with the prescribed criteria, except where a variance has been granted. NYSDEC may grant a waiver from any provision, except for the performance objectives, if necessary for certification of the proposed facility, provided that the facility will meet performance objectives and there will be no significant adverse impacts on the public health, safety, or welfare, the environment, or natural resources.

The NYSLLRW Management Act of 1986 required the Siting Commission to apply to NYSDEC for certification of the selected site and method by December 1, 1988, to allow sufficient time for a disposal facility to be operational by January 1, 1993 (NYSLLRWAC, 1988/b). By the time the Siting Commission and Advisory Committee were both fully operational, however, there was less than a year until the deadline for filing the application.

The Siting Commission initiated site- and method-selection activities in August 1987 when Siting Commission staff began preparing a Request for Proposals (RFP) for assistance in site-selection, method-selection, and preparing the certification application (NYSLLRWSC, 1987). The RFP was distributed in December 1987, and nine complete proposals were received by the due date in February 1988. In March 1988, the Siting Commission announced the selection of Roy F. Weston, Inc., (referred to hereafter as Weston) as contractor (NYSLLRWAC, 1988/b).

During Spring and Summer 1988, the Siting Commission and its contractor developed the Draft Site- and Method-Selection Plans, which called for a three-phased approach to select feasible method-site combinations (NYSLLRWSC, 1988/a,b). In the first phase, selection of sites and methods would be conducted in parallel.   In the second phase, preliminary engineering designs for site/disposal method combinations would be developed using site-specific data.   In the third phase, these combinations would be evaluated to select a preferred site/disposal method. Both method and site selection would be based on multiple-criteria frameworks.   First, exclusionary criteria would be used to narrow the field of potential alternatives, and, then, preference criteria would be applied to evaluate the remaining potential candidates.   A draft environmental impact statement would be prepared with regard to the preferred site(s) and method(s).

Planned public participation programs included a workshop where representatives of State and local government, industry, and interest groups reviewed and commented on the relative weights of the criteria for site and method selection; development and dissemination of public information materials, including a quarterly newsletter and factsheets on related subjects; a series of public meetings; and solicitation of public review and comments on major documents.

In September 1988, the Siting Commission published the Generic Scope of Draft Environmental Impact Statement (DEIS) (NYSLLRWSC, 1988/c). It contained a preliminary list of issues that needed to be addressed in response to the State Environmental Quality Review (SEQR) requirements,[10] plus other topics[11] that should be addressed (NYSLLRWSC, 1988/c).

In October 1988, the Siting Commission held six public information meetings across the State to provide information and to receive public comments on the Draft Plans for Site and Method Selection, the Generic Scope of DEIS, as well as on the Draft Statewide Exclusionary Screening Report (NYSLLRWSC, 1988/a-d), and a summary of the Recommendations for State Assistance to Localities (Working Group, 1988).   Approximately

---

[10] These included a description of the background and need for the disposal facility; a summary of the plans for site- and method-selection; a description of the environmental settings for the preferred site(s) and the alternative candidate site(s); analysis of significant environmental impacts of the preferred and alternative site(s) and method(s); identification of mitigation measures to reduce or eliminate potential negative impacts from the facility; identification of negative impacts which cannot be avoided; analysis of alternatives to the preferred sites(s) and method(s); identification of resources which will be irrevocably committed to the facility; and identification of any aspect of the facility which will stimulate growth in the affected area.

[11] These included recovery and retrieval; mixed wastes at the site; local oversight committees; aid to localities; economic implications of changes in the volumes of LLRW; radioactive biological infectious waste; Below Regulatory Concern LLRW classification; public acceptance; impacts if the facility is not operational by the legislated deadline; and LLRW from decommissioning nuclear power plants.

300 people across the State attended these meetings (NYSLLRWAC, 1989/a). The most frequently expressed public concerns about the design of the process were problems associated with integrating the site- and method-selection processes, perceived limits in the scope of the DEIS, and the inadequacy of public involvement in the process (NYSLLRWSC, 1988/f, pp. 4-17). In November 1988, the Siting Commission issued final plans for selecting one or more sites and disposal methods for LLRW disposal in New York State (NYSLLRWSC, 1988/g,h).

After the announcement by the Siting Commission of candidate areas in December 1988, the process of site and method selection was significantly delayed, as Weston fell behind the schedule specified in the scope of work. Delays in the siting process were regarded as understandable, given the increased levels of public concern and opposition about site selection. Delays in method selection, however, had no readily apparent extenuating circumstance (NYSLLRWSC, 1991/a). In addition, critical comments about the contractor's performance were made in a number of public forums.

As a result of an internal assessment, the Commission decided to request an independent evaluation of Weston's program management capabilities and performance. The request for proposal was published in November 1989, and an audit contract was signed with KPMG Peat Marwick in May 1990. Even before the audit was completed, the Siting Commission announced a number of improvements to its approach to project management (NYSLLRWSC, 1991/a).

An audit report was published by the Siting Commission in February 1991. The report concluded that the project management of Weston was deficient with respect to cost and schedule control (NYSLLRWSC, 1991/a). In September 1991, the Commission terminated its contract with Weston. By December 1991, the Siting Commission had selected four technical support contractors through a competitive bid process.

The NYSLLRW Management Act Amendments of 1990 changed the overall procedure, directing that a preferred disposal method should be chosen by the Siting Commission and conceptually approved by NYSDEC before the site-selection process was initiated, rather than conducting site and method selection simultaneously. The requirement to address method selection before site selection responded to the expressed desires of citizen groups to know what a facility would look like before considering it.

The new law also required that a separate draft generic environmental impact statement accompany the application to NYSDEC for a decision on the preferred disposal method. In addition, the law required that the public be given greater opportunities to review and comment on the site- and method-selection processes. In response, the Siting Commission announced its intention to develop new procedures for method and site selection and communication with the public (NYSLLRWSC, 1991/b). As of December 1992, development of the above procedures was in progress.

SITE SELECTION FOR THE PERMANENT DISPOSAL FACILITY

A chronological summary of major events related to site selection for the permanent disposal facility in New York State is given in Table 8.

The NYSEO report on LLRW management recommended the West Valley site for interim storage, but did not identify a preferred site for permanent disposal. The possibility of reopening the West Valley site for storage emerged as a controversial issue at public hearings and among the NYSEO Advisory Committee (NYSEO, 1984/b, Vol. 1, Attachment 1).

The LLRW Management Act of 1986 (NYS Ch. 673, 1986) prohibited use of the West Valley site. It specified that site selection should consider transportation accidents; ground or surface water contamination; adverse impacts on the natural environment and ecology, water and air quality, and wildlife; effects on scenic, historic, cultural, and recreational values; population density; transportation adequacy; and economic impacts on local governmental units (NYS Ch. 673, 1986).

NYSDEC developed several siting criteria as part of the draft 6 NYCRR Part 382 regulations issued in July 1987 (NYSDEC, 1987/a, Vol. 2). Regardless of the disposal method chosen, any site used for permanent disposal must meet specific requirements grouped under nine headings: general; surface water and hydrology; mineral and exploitable resources; geology and soils; groundwater and hydrogeology; seismic risk; population, land use, and ownership; traffic and transportation; and cultural, recreational, and natural resources. Several additional criteria were specified for above-ground and below-ground disposal methods and for underground mined repositories. In Autumn 1987, NYSDEC conducted a series of public hearings on these draft regulations. Comments at the hearings, as well as written responses, led to a number of modifications, deletions, and additions to the final regulations (NYSDEC, 1987/b).[12]

The Siting Commission was required by 6 NYCRR Part 382 to follow a systematic procedure and define a set of objective criteria for selecting a preferred site. The requirements stated that these criteria should ensure that the selected site would meet the site suitability criteria of Part 382

---

[12] A summary of the complete requirements for permanent disposal sites, as specified by the final version of 6 NYCRR Part 382 Regulations, appears in the Appendix, Table A.1.

Table 8. CHRONOLOGICAL SUMMARY OF KEY EVENTS RELATED TO THE SITE-SELECTION PROCESS

Date	Event
1984	The NYSEO report indicated that there was public opposition to the use of West Valley as the disposal facility site.
1986	The use of West Valley as a LLRW disposal site was prohibited as part of the LLRW Management Act.
1987	NYSDEC developed siting criteria as part of 6 NYCRR Part 382 regulations.
1988	The Siting Commission began implementation of the site selection process. The Siting Commission held a Site- and Method-Selection Workshop to revise criteria and establish weighting factors. The Siting Commission released the Statewide Exclusionary Screening Report. The Siting Commission released the Final Site- and Method-Selection Plans. The Siting Commission released the Candidate Area Identification Report.
1989	Counties containing candidate areas submitted responses to the Siting Commission. The Siting Commission and Department of Health held public meetings in the candidate areas. Opposition organizations were formed to fight the siting process. The Siting Commission released the Report on Potential Site Identification. Governor Cuomo met with opposition groups and officials from Cortland and Allegany counties. The Siting Commission published the draft Generic Site Characterization Plan. Public opposition prevented the Siting Commission from gaining access to sites in Cortland and Allegany counties.
1990	Public opposition prevented the Siting Commission from gaining access to potential sites and operating information offices in Cortland and Allegany counties. Public opposition prevented the Siting Commission from conducting a scheduled walk-over of the potential site in Caneadea, Allegany County. The Town of Ashford lifted its ban on LLRW disposal in the town.
1991	In a non-binding referendum, residents of the Town of Ashford voted against siting a disposal facility at West Valley. The Ashford Town Board unanimously voted to ask the State Legislature to rescind the prohibition on the use of West Valley as the disposal facility site. Ten West Valley residents sued the Ashford Town Board requesting that the Court vacate and annul the resolution of the Town Board.
1992	Lawsuit against the Ashford Town Board was not upheld by State Appellate Court. The U.S. General Accounting Office concluded that the Siting Commission did not exactly follow its procedures.

(NYSLLRWSC, 1988/a).   In response to these requirements, the Siting Commission first identified performance objectives, regulatory requirements, and other environmental, social, policy, and economic objectives.   It then categorized the objectives into general siting factors and developed specific measures for assessing the degree of compliance with each factor. For each measure, the Commission specified the level at which a site would be categorized as most favorable, more favorable, favorable, less favorable, least favorable, or unsuitable.

This process produced a set of 13 siting factors, which were identified as: geology; natural resources; groundwater hydrology; surface water hydrology; meteorology; air quality; ecology; radiology; demography; land use; transportation; socioeconomics; and cultural resources and aesthetics. For each siting factor, Weston and the Siting Commission defined a set of specific criteria. Based on their interpretation of the regulatory requirements, the Siting Commission divided these criteria into preference and exclusionary criteria (NYSLLRWSC, 1988/a).[13]

The Siting Commission held a Site- and Method-Selection Workshop in August 1988 to review and receive public comments on the draft site- and method-selection plans, to establish weighting and scaling factors for site and method selection, and to provide a forum where participants could exchange views and information (NYSLLRWSC, 1988/e). Participants in the workshop included Advisory Committee members, State officials, county officials and representatives of local governments, and representatives from environmental protection groups, among others (NYSLLRWSC, 1988/e).

Workshop participants were asked to assign numeric weights that indicated the relative importance of the criteria. Each individual independently made the weight assignments several times, with intermediate sessions of review and group discussion. Different weights were estimated for above-ground technologies, below-ground technologies, and mine technologies. There was a great deal of variation in the weights that different participants assigned to each criterion.[14] Following the workshop, Weston and Siting Commission staff independently weighted the criteria (NYSLLRWSC, 1988/i). The final set of criteria and weights used in later analyses represented a combination of the weights from the workshop and staff.[15]

Draft site- and method-selection plans were published and distributed in September 1988 (NYSLLRWSC, 1988/h; NYSLLRWSC, 1988/l). In October 1988, the Siting Commission held a series of public meetings, attended by approximately 300 people across the State, to provide information and receive public comments (NYSLLRWAC, 1989/a).

---

[13] The complete list of criteria, as proposed in the Draft Site-Selection Plan, appears in the Appendix, Table A.2 (NYSLLRWSC, 1988/a).

[14] Summary statistics of the site-selection weights appear in the Appendix, Table A.3 (NYSLLRWSC, 1988/e).

[15] The weights assigned by workshop participants, by staff members, and the final "base case" weights used in later analyses appear in the Appendix, Table A.4.

A number of changes were made from the draft before the final site-selection plan was issued. In the final plan, the Siting Commission defined many criteria in ways that provided readily measurable, mappable indicators for identifying candidate sites (NYSLLRWSC, 1988/h). The final plan noted that different weighting factors might be applied at different steps of the process (NYSLLRWSC, 1988/h). The final site-selection plan specified 61 criteria.[16]

In November 1988, the Siting Commission issued the final site-and method-selection plans, which specified that method and site selection would be carried out in parallel and would be integrated to conduct an environmental impact assessment. In January 1989, after several potential volunteer sites had gained public attention, the Siting Commission modified the site-selection plan to specify that volunteer sites must be at least as good as the sites being considered at that stage (NYSLLRWSC, 1991/b).

In the first step of the site-selection process, a subset of the exclusionary criteria that could be easily and accurately identified were applied.[17] This step excluded approximately 30 percent (or 9.7 million acres) of the State (NYSLLRWSC, 1988/i). The first exclusionary map, the Statewide Exclusionary Screening Report, was published in October 1988.

To screen the remaining 70 percent of the State, a number of exclusionary criteria and selected preference criteria were added (NYSLLRWSC, 1988/i).[18] Data were obtained from or verified by NYSDEC, the Energy Authority, and the National Park Service. Data processing was conducted with a Geographic Information System (GIS) developed by Weston (NYSLLRWAC, 1989/a). Overlay maps were created that displayed the level of compliance with each criterion throughout the State and helped to identify potential candidate areas (NYSLLRWSC, 1988/i).

The additional exclusionary criteria increased the total area of the State excluded to 40 percent. The preference criteria were then applied to the remaining 60 percent of the State. Favorability scores were assigned to each parcel of land within each area for all criterion. Total scores -- the sums of the products of the weighting and scaling factors for all the criteria -- were then computed. The Siting Commission chose a cutoff score that eliminated all but approximately five percent of the State. From composite preference maps, 30 potential candidate areas were identified.[19]

---

[16] The 61 criteria contained in the final selection plan, along with the scales for measuring them, appear in the Appendix, Table A.5.

[17] These included the criteria relating to primary aquifers and the Long Island aquifer; towns and cities with a population density of more than 1,000 persons per square mile; the Adirondack Park; the Catskill Park; other State parks; and Indian reservations.

[18] Criteria used at this stage are listed in the Appendix, Table A.6.

[19] The candidate areas and their performance scores are listed in the Appendix, Table A.7.

Sensitivity analyses were conducted to evaluate the effect of changing the weighting scheme used to evaluate potential candidate areas. The results indicated that the scores of potential candidate areas were not very sensitive to changes in the weighting scheme.

Existing and potential new mines were considered separately. Twenty-five existing mines with high scores on the evaluation criteria were selected for further consideration (NYSLLRWSC, 1988/i).

To select a smaller number of areas having the greatest potential for containing suitable sites for a LLRW disposal facility, the Siting Commission qualitatively examined each candidate area or mine, considering additional criteria.[20] Assessments were based on readily available data sources, and the quality of the data was uneven (NYSLLRWSC, 1988/i). In the final step of the candidate area identification process, the Siting Commission gave additional weight to the potential for construction of new mines, proximity to generators, and proximity to other candidate areas.

In December 1988, the Siting Commission announced the selection of ten candidate areas to be considered further during the next stage of the site-selection process; these areas were located in Allegany, Cayuga, Chenango (2), Clinton, Cortland, Montgomery, Orange/Ulster, Oswego, and Washington counties (NYSLLRWSC, 1988/i).

To handle the increase in public participation following the announcement of the candidate areas, the Siting Commission assigned each candidate area a staff liaison person to be a point of contact, answer questions, and seek input in selection of potential sites. In December 1988, a toll-free telephone information line was established (NYSLLRWSC, 1989/f).

The amount of public interest and concern grew dramatically following publication of the Candidate Area Identification Report. Public meetings, held in the ten candidate areas in January 1989, attracted approximately 15,000 attendees (NYSLLRWAC, 1989/a). Numerous citizens stated that they did not want the disposal facility in their areas and raised questions about the type of waste to be disposed of at the facility; the health effects of LLRW; potential effects on surrounding property values; the regulatory framework; transportation risks; fairness of siting near to or distant from sources of the waste; the process of purchasing the selected site; projected impacts on the surrounding communities; offsetting benefits planned for the host community; and plans for local oversight (NYSLLRWSC, 1989/h).

Before the date of the scheduled public hearing, the Siting Commission requested that counties, local officials, and private citizens provide technical

---

[20] These included stratigraphic complexity; subsurface dissolution; State reforestation areas; proximity to operating mines; well head areas; wetlands; drainage and runoff; erosion; upstream impoundments; surface water quality; nearby sources of radioactive releases; distance from highly populated places; prime farmlands; proximity to major highways; and archaeological sites, plus several additional criteria for the mines (NYSLLRWSC, 1988/i).

information pertaining to the site-selection criteria for each candidate area. This information would assist in eliminating potential sites within the candidate areas.

In late January 1989, counties with candidate areas submitted responses to the Candidate Area Identification Report. In February 1989, several county legislatures passed resolutions opposing the siting of a LLRW facility or protesting the process itself; some towns, villages, and school boards did the same.

In late February and early March 1989, the New York State Department of Health (NYSDOH) held public meetings in each candidate area to present information on LLRW, radiation, and health. Additional public meetings were held in the candidate areas by the Advisory Committee. Recommendations by members of the public included a phased shut-down of nuclear power plants, taxation of the LLRW disposal facility at full value, appointment of the Host Area Local Advisory Committee by local officials, and specification of compensation packages before an interim storage site is selected (NYSLLRWAC, 1989/b). Some people were dissatisfied with the way the meetings were conducted and expressed feelings of powerlessness on the part of the candidate areas (NYSLLRWAC, 1989/b).

Following the announcement of the candidate areas, a number of opposition organizations formed in the affected counties. For example, the Citizens Against Radioactive Dumping (CARD) formed in Cortland and Chenango counties and, in January 1989, took the positions that a LLRW disposal facility should not be located in Chenango and Cortland counties, in agricultural, food-producing areas, or in an area in which the majority of drinking water is supplied by groundwater sources; town or county governments should not be overridden; LLRW should be treated close to the areas of greatest production; Class C waste should be classified as High-Level Radioactive Waste; and radioactive waste should be reduced by all generators (Citizens Against Radioactive Dumping, 1989).

The next step of the site-selection process involved identifying a small number of potential disposal sites within the ten candidate areas. Using a set of criteria that addressed readily mappable conditions, each candidate area was screened using the GIS. Limited on-site inspections were performed at each site to identify readily apparent unfavorable siting conditions and to select a smaller number of sites for further consideration. These potential sites were then evaluated or reevaluated applying the siting criteria to site-specific information and information provided by local officials. The sites were compared to select those appropriate for pre-characterization[21] (NYSLLRWSC, 1989/d).

---

[21] Pre-characterization involves background data search, field reconnaissance, and limited subsurface investigations (NYSLLRWSC, 1989/e). Its purpose is to identify the most

To develop composite favorability maps that eliminated some sites from further consideration, the Siting Commission used 12 exclusionary criteria.[22] The Siting Commission then used 27 selected preference criteria to evaluate the relative favorability of the remaining sites (NYSLLRWSC, 1989/d).

Ninety-six specific sites within the ten candidate areas were identified and assessed, and the list was narrowed to 51 sites (NYSLLRWSC, 1989/d). The qualitative information used at this stage included proximity to incompatible structures (i.e., schools, hospitals, churches, commercial centers, or nursing homes), soil types and agricultural activity, drainage, slopes, and transportation access.

Limited site inspections were performed at each of the 51 sites, plus four volunteer sites that had come to the Commission's attention. The site surveys helped identify unanticipated but readily apparent unfavorable conditions and led to selecting, for further consideration, 19 sites in Allegany, Chenango, Cortland, Montgomery, and Washington counties. Finally, the 19 sites were compared on the basis of the siting criteria and other technical considerations.[23]

In September 1989, the Siting Commission's Report on Potential Sites Identification (ROPSI) specified the five remaining potential sites; two were in Cortland County, and three were in Allegany County.[24] In September 1989, the Siting Commission voted to initiate precharacterization studies, including those requiring on-site access (NYSLLRWAC, 1989/c).

Substantial controversy surrounded the announcement of the five potential sites. Both Cortland and Allegany counties responded with detailed comments contesting the suitability of those sites and the appropriateness of the siting process. Cortland County challenged numerous alleged discrepancies between the Plan for Selecting Sites for Disposal of Low-Level Radioactive Waste and the actual site-selection process as implemented

---

promising candidate sites for detailed characterization and to discover any "fatal flaw" that would eliminate sites from further consideration.

[22] These included State reforestation areas; primary and principal aquifers; surface water bodies; 100-year floodplains; NYSDEC-regulated wetlands; areas not in attainment of air pollution standards; habitats for threatened or endangered species; population density greater than 1,000 persons/square mile; federal protected lands; State protected lands; mineral soil groups 1-4; and historic places (NYSLLRWSC, 1989/d).

[23] These included land ownership, nearby activities, future development, agricultural soil quality, size and shape, disposal method flexibility, and geographic location (NYSLLRWSC, 1989/d).

[24] The *Caneadea Site* consists of 1,039 acres and is located in Allegany County in the eastern portion of the Town of Caneadea and the western portion of the Town of Allen. The *Allen Site* consists of 780 acres in the southeastern corner of the Town of Allen in Allegany County. The *West Almond Site* consists of 918 acres in the southwestern corner of the Town of West Almond, Allegany County. The *Taylor North Site* consists of 686 acres in the northern part of the Town of Taylor, Cortland County; it's owner volunteered to make the land available for a disposal facility site. The *Taylor Central Site* consists of 473 acres in the central part of the Town of Taylor, Cortland County (NYSLLRWSC, 1989/g).

(Monaco, 1989). The response by Allegany County described alleged misapplication of criteria, claimed that the Siting Commission's responses to concerns submitted previously had been inadequate, and argued that the Commission siting process violated applicable law and regulations.

In November 1989, at a public hearing attended by approximately 5,000 people, representatives of Cortland County reiterated and extended their criticisms of the Commission's methodology, forthcomingness, responsiveness, and technical performance (Monaco, 1989). Several lawsuits were filed against the Siting Commission by opposition groups, and Cortland County sued the Siting Commission regarding restrictions on the use of funds appropriated by the State to the County.

In December 1989, the Siting Commission published a draft Generic Site Characterization Plan describing requirements for characterizing a potential site in any candidate area. This plan was to be supplemented by Site-Specific Characterization Plans for each potential candidate site. According to the draft, the objectives of characterization were to provide the information needed to assess the relative suitability of the candidate sites and to prepare documents required for certification of the preferred site(s), including the Draft Environmental Impact Statement (DEIS), the Safety Analysis Report (SAR), and Preliminary Design of the Facility (NYSLLRWSC, 1989/e). Characterization of candidate sites was to be divided into three progressively more detailed stages: the exploration, confirmation, and preoperational monitoring stages (NYSLLRWSC, 1989/e). Pre-characterization would be conducted to assist in identifying the most promising candidate sites for detailed characterization and eliminating other sites from further consideration.

The Commission attempted to proceed with on-site precharacterization of the five sites. From December 1989 to April 1990, however, Siting Commission staff were repeatedly prevented by protestors from conducting "walk-overs" of the sites in Cortland and Allegany Counties (*Olean Times Herald*, December 14, 1989). There also were several confrontations between State representatives and area residents when the Siting Commission tried to open information offices in the two counties (*Olean Times Herald*, March 11, 1990).

In April 1990, the Siting Commission scheduled a walk-over of the potential site in Caneadea, Allegany County. This led to the arrest of 39 protestors who blocked access to the site, the removal of several barricades, and an encounter with masked protestors on horseback (*Olean Times Herald*, April 6, 1990). After this incident, the Governor recommended that the Siting Commission halt on-site, precharacterization activities and concentrate instead on other disposal facility development work.

In August 1990, the Town Board in Ashford, where the Western New York Nuclear Service Center is located, voted to lift its five-year old restriction on LLRW shipments to the Town. Shortly thereafter, some Ashford residents formed a group to oppose LLRW being shipped into the community (*Olean Times Herald*, August 21, 1990).

In June 1991, the West Valley Chamber of Commerce President announced that a five-member committee supported the location of the LLRW disposal facility at the Western New York Nuclear Service Center in exchange for a $4.23 million community benefits package negotiated by representatives of the Energy Association, a private organization representing the interests of the State's investor-owned electric utility companies (*Buffalo News*, June 13, 1991). Local reactions to the proposed benefits package were mixed (*Buffalo News*, June 14, 1991). The Siting Commission emphasized that it was uninvolved in the negotiations, and the Governor emphasized that the West Valley proposal was not a State-sponsored initiative (*Olean Times Herald*, June 20, 1991).

A public meeting, attended by approximately 600 people, was held in West Valley in late June 1991. Discussion concerned the proposal to bring a LLRW disposal facility to West Valley. Opponents of the proposal argued that the site was unsuitable, that federal funding for current cleanup efforts would be jeopardized, and that secrecy surrounding the development of the proposal was inappropriate (*Buffalo News*, June 21, 1991).

A bill intended to incorporate the results of the negotiations between the Town and the LLRW generators was introduced in the New York State Senate and Assembly. The bill authorized and directed the Energy Authority to construct and operate LLRW management or disposal facilities in the Town of Ashford at the Western New York Nuclear Service Center and specified host area benefits and rights (NYS Senate Bill S.6283-A, Assembly Bill A.8748-A, 1991).

In late June 1991, the Ashford Town Board voted 4-1 to table a decision regarding acceptance of the benefits package. In early July, the Town Board appointed a citizens' committee to obtain further information about the proposal to bring the LLRW disposal facility to West Valley (*Olean Times Herald*, July 2, 1991). Meanwhile, U.S. Representative Bill Paxon stated his opposition to a LLRW facility at West Valley. A nonpartisan citizens' group sponsored an informational meeting where representatives from generators, citizen groups, and State agencies were available to respond to questions.

In early July 1991, a non-binding referendum of residents and non-resident property owners was held on the question of whether the Town of Ashford should consent to siting a LLRW management facility on State-owned land in the Town, provided the site was determined to be suitable, that the health, safety and welfare of the area would be adequately protected, and that benefits and compensation were paid as described in the Senate and Assembly bills. The referendum was defeated, but the Town Board unanimously voted to ask the State Legislature to rescind the prohibition on the use of West Valley as the LLRW disposal site and to ratify the compensation package, contingent on several modifications (Town of Ashford, 1991), which were subsequently made to the proposed legislation (NYS Senate Bill S.6283-A, Assembly Bill A.8748-A, 1991).

In late July, a group of West Valley residents brought suit against the Ashford Town Board seeking to overturn the Board decision to accept a LLRW disposal facility in exchange for the $4.23 million benefits package; the petition against the Town was upheld. In Autumn 1991, the Town appealed the decision, and in Summer 1992, the Appellate Division of State Supreme Court overturned it.

In mid-1992, the Siting Commission implemented a new Geographic Information System (GIS) which is accessible for the public to simulate the site-selection process. The Commission also initiated an Excluded Area Study to assess the impact of all federal and state rules and regulations which affected the siting process.

At the request of U.S. Senator Alfonse M. D'Amato, the U.S. General Accounting Office (GAO) conducted an inquiry into the site-selection process. In August 1992, GAO released its report, which concluded that the Siting Commission did not strictly follow its own siting procedure with regard to the Taylor North site. The GAO report concluded that the site contained more productive agricultural land than State regulations permit for LLRW facilities and that the numerical score assigned to the site was less than the minimum score established by the Commission as a threshold for further consideration (GAO, 1992).

PERMANENT DISPOSAL METHOD SELECTION

A chronological summary of major events related to permanent disposal method selection in New York State is given in Table 9.

The general framework for disposal method selection was created by the NRC in the 10 Code of Federal Regulations (CFR) Part 61 regulations, which establishes minimum requirements for near-surface land disposal methods. The Part 61 regulations specify performance objectives and define facility design criteria.[25] The regulations require the operator to observe, monitor, and carry out necessary maintenance and repairs at the disposal site

---

[25] These include long-term isolation and avoidance of the need for active maintenance after site closure; compatibility with the site closure and stabilization plan; complementing or improving the ability of the disposal site's natural characteristics; minimizing water infiltration; minimizing the contact of water with waste; the capability to be characterized, analyzed, and monitored; and the capability to be remedied.

Table 9. CHRONOLOGICAL SUMMARY OF KEY EVENTS RELATED
TO PERMANENT DISPOSAL METHOD SELECTION

Date	Event
1984	The NYSEO report recommended rejecting deep mine disposal but continuing further study of shallow land burial, mined cavity disposal, and engineered, monitored disposal.
1986	The LLRW Management Act prohibited use of shallow land burial.
1987	The NYSDEC developed method-selection criteria as part of 6 NYCRR Part 382 regulations.
1988	The Siting Commission began implementation of the method-selection process. The Siting Commission held a Site- and Method-Selection Workshop to revise criteria and establish weighting factors. The Siting Commission released Final Site- and Method-Selection Plans.
1989	The Siting Commission issued the Disposal Method Screening Report.
1991	The Siting Commission issued a disposal method-selection plan.

until the license is transferred to the disposal site owner. Responsibility for
the disposal site must be maintained by the licensee for at least five years.
The regulations require institutional control of access to the site for up to 100
years.

The 1984 draft NYSEO study reviewed five permanent disposal methods
(NYSEO, 1984/a): shallow land burial, intermediate depth burial, deep mine
disposal, offshore disposal, and engineered, monitored disposal.   These
methods were analyzed in terms of the performance objectives and facility
design criteria prescribed by the NRC Part 61 regulations, as well as
considerations of cost, time, legal context, and feasibility. The draft report
recommended further evaluating shallow land burial, intermediate depth
burial, and engineered, monitored disposal. It rejected deep mine disposal
and offshore disposal because of concerns about costs, safety, capacity,
feasibility, and legality.

At public hearings, participants raised concerns about shallow land burial
because of previous problems with this technology. A number of citizens
and elected officials suggested that above-ground, engineered, monitored,
retrievable storage should be further investigated (NYSEO, 1984/b). Some
members of the Advisory Committee recommended considering engineered,
monitored, shallow-land burial, but a minority group within the Committee
strongly opposed this recommendation (NYSEO, 1984/b, Vol. 1.,
Attachment 1). The final NYSEO report recommended further evaluating
shallow land burial, intermediate land burial, mined cavity disposal, and
engineered, monitored disposal.

The NYSLLRW Management Act of 1986 prohibited shallow land burial.
It directed NYSDEC to promulgate regulations applicable to above-ground

monitored disposal, underground mined repository disposal, and other disposal methods except shallow land burial. It specified that method selection should consider probable impacts on the natural environment and ecology; scenic, historic, cultural, and recreational values; water and air quality; and wildlife. The process should also consider the potential to avoid or mitigate the harm from unanticipated releases of LLRW or contaminated materials and the ability to retrieve or recover the waste. The facility was required to have the capacity to accommodate at least 30 years of New York State LLRW.

Before promulgating regulations for method selection, NYSDEC investigated above-ground vault disposal, below-ground vault disposal, underground mined repository, earth-mounded concrete bunkers, boreholes, injection, abandoned mines, ocean disposal, and space disposal (NYSDEC, 1987/a). NYSDEC rejected injection, abandoned mines, ocean disposal, and space disposal as infeasible or inappropriate, and rated all other alternative methods as feasible. Part 382 regulations were, therefore, developed so that they would be applicable to above-ground, below-ground, and underground mined disposal.

The Part 382 draft regulations were based on the prescriptions of the NRC Part 61 regulations, the New York State LLRW Management Act, and all other applicable federal and State laws and regulations. NYSDEC set lower permissible levels of direct gamma radiation and whole body radiation than those specified in the federal regulations and enumerated a number of further requirements.[26]

Written comments on the draft regulations, as well as comments raised at public hearings, reflected disagreements about the advantages and disadvantages of the various methods, the validity of the performance objectives, and the importance of certain proposed method-selection criteria. Environmental groups generally favored above-ground, non-buried facilities which are fully recoverable, retrievable, and monitored during the full hazardous life of the waste (NYSDEC, 1987/b, Vol. 2, E4, E5). Other groups claimed that assuring recoverability would add risk of radiation escape (NYSDEC, 1987/b, Vol. 2, E7). Some organizations criticized NYSDEC for adding more performance objectives and more stringent

---

[26] These included requirements that construction methods and techniques must not adversely affect the facility's ability to meet performance objectives; mixed hazardous and LLRW wastes must not be disposed of unless the facility complied with existing law concerning hazardous waste disposal; the performance of the disposal units must be capable of being characterized, modeled, analyzed, and evaluated; it should be feasible to recover waste during the institutional control period; the facility should provide for long-term isolation and avoid the need for continuing active maintenance; the facility should possess long-term structural stability; and any new access route(s) to the site must be designed and constructed using accepted standards and technology. Several additional criteria were specified for above-ground, below-ground, and mined disposal.

criteria to the NRC performance objectives (NYSDEC, 1987/b, Vol. 2, B6, E8). Conversely, some environmental groups and individuals criticized the NRC performance objectives for being insufficiently strict. The final regulations on disposal methods were issued in December 1987 (NYSDEC, 1987/b), with a few deletions, modifications, and additions to the draft version.[27]

The Siting Commission initiated site- and method-selection activities in March 1988. With the assistance of Weston, the Siting Commission staff screened disposal method alternatives between July 1988 and July 1989. A comprehensive list was developed of disposal method alternatives that had been used or proposed worldwide, and the Commission developed a generic approach to describe all basic land disposal methods in terms of geologic placement and the configuration of barriers or other containment methods.

Five basic types of geologic placement decisions were identified: surface only; surface mound; shallow excavation; tunnel; and, deep/drift mine. Five basic types of barrier or containment configurations were identified: vaults; modular disposal units (MDU); monolith; lined disposal units; and, no barrier. As seen in Figure 2, the combination of the five options for geologic placement and the five barrier or containment configurations produced a matrix that defined 25 possible disposal methods.

Next, the Siting Commission excluded the methods that did not meet regulatory requirements: shallow excavation/no barrier, (i.e., shallow land burial); surface only liner; surface only/no barrier (i.e., unimproved above-ground disposal); surface mound/no barrier; and, tunnel without barrier (i.e., shallow tunnel).

Following exclusionary screening, the Siting Commission initiated a multi-attribute evaluation process. The Siting Commission planned to: identify the factors that discriminate among disposal methods; identify criteria that describe the attributes of each factor; develop a scheme for scoring degree of compliance with the criteria; assign weighting factors reflecting the Siting Commission's judgment of the relative importance of each criterion; assign raw scores to each disposal method based on the Siting Commission's judgment of its degree of compliance with each criterion; multiply the raw scores by the appropriate weighting factor to obtain a weighted score for each criterion; develop an overall weighted score for each disposal method; and, compare the overall weighted scores for the disposal methods (NYSLLRWSC, 1988/g). In July 1988, the Siting Commission identified 11 relevant factors for method selection, based primarily on the

---

[27] These are summarized in the Appendix, Table A.8.

Figure 2. 25 POSSIBLE DISPOSAL METHODS

GEOLOGIC PLACEMENT \ ENGINEERED BARRIER	VAULT	MDU	MONOLITH	LINER	NO BARRIER
SURFACE ONLY	SURFACE ONLY VAULT	SURFACE ONLY MDU	SURFACE ONLY MONOLITH	SURFACE ONLY LINER	SURFACE ONLY NO/BARRIER
SURFACE MOUND	SURFACE MOUND VAULT	SURFACE MOUND MDU	SURFACE MOUND MONOLITH	SURFACE MOUND LINER	SURFACE MOUND NO/BARRIER
SHALLOW EXCAVATION	SHALLOW EXCAVATION VAULT	SHALLOW EXCAVATION MDU	SHALLOW EXCAVATION MONOLITH	SHALLOW EXCAVATION WITH LINER (SHAFT)	SHALLOW EXCAVATION /NO BARRIER
TUNNEL	VAULT IN A TUNNEL	MDU IN A TUNNEL	SURFACE MOUND MONOLITH	LINED TUNNEL	TUNNEL WITHOUT BARRIER
DEEP/DRIFT MINE	VAULT IN A MINE	MDU IN A MINE	MONOLITH IN A MINE	LINED MINE	MINED REPOSITORY

Source: NYSLLRWSC, 1989/c

NYSDEC regulations.[28]   Next, specific criteria were defined for each factor.[29] For each criterion, an ordinal scale would be used to rate the degree to which the various methods satisfied it. The scale ranged in value from one to five, with one defined as least favorable and five as the most favorable level.

In August 1988, the Siting Commission organized a three-day workshop to review and receive comments on draft procedures, to revise the site- and method-selection criteria, and to give them weights.   Twenty workshop participants, including members of the Advisory Committee, State agencies, citizen and environmental groups, and county representatives, completed the weighting exercises. There was a great deal of variation in the weights that different participants assigned to each criterion.[30] Following the workshop, Weston and the Siting Commission staff independently weighted the criteria and revised the initial list of criteria based on the comments from the workshop discussions.[31]

In public meetings in October 1988, the most frequently-raised issues about the method-selection criteria concerned waste retrievability, recoverability, the ease and effectiveness of remedial action, and the importance of surveillance and monitoring (NYSLLRWSC, 1988/f). The Siting Commission did not change the criteria and weights after this series of public meetings.

In November 1988, the Siting Commission published the Plan for Selecting Methods (NYSLLRWSC, 1988/g), which described the steps already taken and further detailed future steps of method selection. The plan prescribed that the method-selection process should be based on multi-attribute numerical evaluation and ranking, using the factors defined in the earlier steps, but allowed that the criteria, weights, and scaling measures might vary at different times in the process. The final method-selection plan presented 13 factors and 23 criteria (modified from the criteria presented at the workshop and the draft plan) to be used in the first step of method selection.[32]

---

[28] These included public safety; worker safety; inadvertent intruder safety; long-term effectiveness of barrier systems; impact on local community; economics and costs; licensability; method practicality; non-radiological environmental impacts; operational flexibility; and, efficiency of mitigative measures (NYSLLRWSC, 1988/g).

[29] The initial list of criteria is reproduced in the Appendix, Table A.9 (NYSLLRWSC, 1988/b).

[30] Summary statistics of the method-selection weights are given in the Appendix, Table A.10 (NYSLLRWSC, 1988/e).

[31] The final set of 23 criteria and the adjusted weights assigned by workshop participants (on average), Commission staff members, and the combination of these weights, called "base case," are given in the Appendix, Table A.11.

[32] These are given in the Appendix, Table A.12.

The Siting Commission next assigned raw scores based on the degree to which each of the disposal methods that had been identified during preliminary screening was judged to satisfy each criterion. Weighted scores for each method were calculated using the base case weights. Evaluation scores were calculated separately for Class A wastes and for Class B and C wastes.[33] Because of the similar characteristics of many of the disposal methods, the evaluation produced relatively similar overall scores for many methods.

The Siting Commission then used several additional criteria to narrow the choices.[34] The Siting Commission decided to retain a spectrum of above-ground, below-ground, and underground mined repository facilities and to analyze a variety of method types, rather than multiple variations of a few methods. The Siting Commission noted that a number of design studies by other states and compacts had concluded that concrete vaults and modular disposal units were superior. The Commission found that the preceding considerations were still insufficient and inconclusive for selecting methods for conceptual design, so the methods were re-evaluated with respect to several additional factors.[35]

Based on this assessment, several options were selected for conceptual design: a generic or standard vault that could be placed either on or below the ground surface and that would be covered with earthen material; a generic or standard modular disposal unit (MDU) that could be placed either on or below the surface and that would be covered with earthen material; a generic or standard lined disposal unit, that could be placed either on or below the ground surface and that would be covered with earthen material; and, a generic or standard lined tunnel/drift mine.

Because the evaluation indicated that different methods were likely to be superior for different classes of waste, the Siting Commission concluded that different concepts might be used at the same site to handle different classes of waste and that basic methods might be combined to provide multiple engineered barriers (NYSLLRWSC, 1989/c, ES-10). Based on further analysis, the Siting Commission staff and Weston concluded that conceptual designs should be developed for: earthen-covered vaults for Class A waste; earthen covered vaults for Classes B and C waste; earthen-covered modular disposal units for Class A waste; earthen-covered lined vertical shafts for Class B waste; and a drift mine for Classes A, B, and C waste (NYSLLRWSC, 1990/a,b).

---

[33] These are given in the Appendix, Table A.13.

[34] These included regulatory considerations; public preferences; efficiency of the design effort; experience of other states and regional compacts; and, results of special studies performed by New York State (NYSLLRWSC, 1989/c).

[35] These included retrievability, minimal active maintenance requirements, reliability of water management systems, and confidence in the long-term durability and performance of barriers (NYSLLRWSC, 1989/c).

In 1989, NYSDOH disseminated Statewide a brochure on disposal technologies (NYSDOH, 1989/b). In Spring 1990, NYSDOH sponsored several public seminars on drift mines and engineered vaults (NYSDOH, 1991/c).

According to the plans, a conceptual design package would show general layout of structures, describe general design and operational features, provide cost estimates, identify site characteristics important to method, and include performance assessments of each method (NYSLLRWSC, 1990/b). Conceptual design started after potential sites had been identified in September 1989. It was partially completed, and draft analyses were submitted by the contractor, when work was suspended as a result of the passage of the amendments to the LLRW Management Act in July 1990. In the evaluation of the Siting Commission staff, the method-selection work done by the contractor to that date was substantially incomplete, and several gaps existed in the data expected from the conceptual design effort itself (NYSLLRWSC, 1990/b).

Most public protest focused on site selection, but the method of disposal became an issue for many of those who opposed the selection of one of the candidate sites. Many people in candidate areas felt it was unfair to ask them to consider siting a facility when they didn't even know what the facility would be.

In July 1990, the LLRW Management Policy Amendments Act (NYS Ch. 913, 1990) was passed. The amendments require selecting a preferred disposal method and conceptually approving it before the site-selection process continues.

The Siting Commission developed plans and management documents in order to comply with the amendments. In January 1991, it issued a disposal method-selection program (NYSLLRWSC, 1991/b) indicating that the method-selection study was expected to resume in Autumn 1991.

Some environmental groups continued criticizing the method-selection process. In November 1991, a report published by the Cortland County LLRW Office (Snyder et al., 1991) questioned the appropriateness of the engineered disposal methods selected by the Siting Commission and suggested that deep mine repositories should be considered as a potential method.

In September 1992, a project plan on Performance Assessment was prepared by the Siting Commission and sent to technical support contractors for proposals. The project will establish the assessment plan and associated procedures to be used in the study of methods. In addition, the Siting Commission contracted GCI Group, Inc., and Rogers and Associates to provide technical support on public participation for method selection.

COMPENSATION AND INCENTIVES

A chronological summary of major events related to compensation and incentives in New York State is given in Table 10.

Table 10. CHRONOLOGICAL SUMMARY OF KEY EVENTS RELATED TO COMPENSATION AND INCENTIVES

Date	Event
1980	The National Governor's Association (NGA) recommended establishing a package of compensation and incentives for host states and communities.
1984	The NYSEO report concluded that some type of compensation or incentives for the host community would be necessary.
1986	The LLRW Management Act required the NYSDEC to prepare a report on possible aid or assistance to host communities.
1987	NYSDEC released the Assistance to Localities Report.
1988	The Siting Commission chaired a working group to develop a package of local assistance incentives. The working group's recommendations were sent to the Governor.
1989	A bill to formalize a community assistance package was introduced, but no formal action was taken.
1990	The LLRW Management Act Amendments required the Siting Commission to recommend a site-specific mitigation program as part of the environmental impact statement.
1991	A bill to formalize a community assistance package was introduced, but no formal action was taken. Representatives of the generators began to negotiate a community benefits package with West Valley Chamber of Commerce. The West Valley Chamber of Commerce President announced support for locating the LLRW disposal facility at West Valley in exchange for a $4.23 million community benefits package.

The 1984 NYSEO report on LLRW management concluded that hosting a LLRW disposal facility in New York State would likely require some types of compensation or incentives for the community in which the site would be located in order to ensure that the benefits of the facility would outweigh the real or perceived costs and impacts (NYSEO, 1984/b, Vol. 1.).

The 1986 NYSLLRW Management Act dictated that the State pay annual sums in lieu of taxes to the affected municipality or taxing district. It also required that NYSDEC prepare a study to: describe the probable impacts on local governmental units from the presence, construction, operation, maintenance, closure, and post-closure care of a LLRW disposal facility; describe the possible forms or kinds of aid or assistance which might be appropriate to mitigate or provide offsetting benefits; and make recommendations to the Governor and the Legislature about the appropriate forms of State assistance to local communities.

NYSDEC investigated the impacts of a LLRW disposal facility on quality of life, health and safety, public services and finances, residential property values, and socioeconomic factors, such as employment, tourism, and horticulture. NYSDEC reviewed offsets proposed or adopted elsewhere in the United States. NYSDEC defined offsets to include both monetary and non-monetary measures such as public participation in the siting and operation of the facility. Offsets included mitigation of avoidable impacts, compensation for unavoidable impacts, and incentives offering net benefits to a host community. The NYSDEC report stressed that risks should be avoided rather than offset and that any program of compensation and incentives should avoid eliciting, through extensive incentives, a host community volunteer that does not meet the most stringent environmental criteria (NYSDEC, 1987/c).

NYSDEC recommendations included: financial assistance to counties that are short-listed for site selection to ensure their participation in the siting review; formation of a Host Area LLRW Commission (HALC) to monitor the facility's operation; special payments to reimburse the host community for unquantifiable quality of life changes; assignment of ultimate financial liability to the State of New York; reimbursement for net in-migration and facility-generated public services; real property value guarantees; and preferential treatment for local workers and vendors (NYSDEC, 1987/c).

NYSDEC estimated the costs of this offset package at $1,790,000 for the first five years and $7,500,000 for the next 30 years ($250,000 per year). The package was to be funded from fees collected from LLRW generators. According to NYSDEC estimates, the costs of the package would be less than 10 percent of the expected gross revenue of the facility over its life cycle (NYSDEC, 1987/c).

In May 1988, a working group was established and chaired by the Siting Commission to further review, on the basis of the NYSDEC report, the issue of local assistance initiatives for communities affected by a LLRW facility. Representatives from the Siting Commission, the Advisory Committee, NYSDEC, the Energy Authority, NYSDOH, and the Division of Budget participated. The working group reviewed the NYSDEC report and refined the recommendations (Working Group, 1988). In October 1988, the Siting Commission sent the recommendations to the Governor (NYSLLRWSC, 1988/m).

No legislation on a compensation package was introduced until the 1989 legislative session. After the Siting Commission announced candidate areas, the process had became very controversial. In June 1989, an amendment to the NYSLLRW Management Act (New York State Assembly, 1989), largely identical to the measures proposed by NYSDEC, the working group, or both, was introduced in the Assembly, but no action was taken. The amendments of the NYSLLRW Management Act, which passed in July 1990, included an offset element, however. As part of the draft environmental impact statement, the law required the Siting Commission to recommend a site-specific mitigation program that would offset adverse economic, social, and environmental impacts and serve as a form of compensation to the affected areas (NYS Ch. 913, 1990).

In March 1991, a multi-sponsored bill on granting aid to host area communities was introduced in the Committee on Environmental Conservation (NYS, S.3486/A.5487, 1991). The bill was not passed by the Committee, but in July 1991, the Aid to Localities Budget Bill for 1991-92 appropriated $625,000 to NYSDEC for similar grants to both Cortland and Allegany counties.

In April 1991, the Energy Association, a special interest organization representing the State's investor-owned electric utility companies, hired a representative to negotiate a potential host community benefits package with members of the Town of Ashford Chamber of Commerce. In June 1991, the Chamber of Commerce President announced that a five-member committee supported locating a LLRW disposal facility at the Western New York Nuclear Service Center in exchange for a $4.23 million community benefits package (*Buffalo News*, June 13, 1991). In late June 1991, a bill intended to incorporate the results of the negotiations between the Town and LLRW generators was introduced in the Senate and Assembly.

In July 1991, U.S. Representative Bill Paxon announced his opposition to a LLRW facility at West Valley, stating, among other points, that the benefits package offered to the Town was much too small (*Congressman Bill Paxon News*, July 2, 1991). Also in July, Cattaraugus County legislators voted against a proposed resolution endorsing the Ashford Town Board's approval of the bill on a LLRW disposal facility at West Valley and requesting the County's inclusion in the benefits package (*Buffalo News*, July 18, 1991). The Energy Association indicated that generators would still honor their offer of a benefit package to the County despite the vote (*Albany Times Union*, July 18, 1991). In September 1991, Cattaraugus County legislators resolved to support the idea of an economic benefits package for the County if a LLRW disposal facility were constructed in West Valley. The legislators noted that the County has a law banning a LLRW facility in West Valley, while recognizing that a facility might still be placed there if the County law is preempted by State law. As of December 1992, the State Legislature had taken no action regarding the ban.

INTERIM WASTE MANAGEMENT

A chronological summary of major events related to interim waste management in New York State is given in Table 11.

Table 11. CHRONOLOGICAL SUMMARY OF KEY EVENTS RELATED TO INTERIM WASTE MANAGEMENT

Date	Event
1983	NYSLLWG called for an in-State interim disposal capacity.
1984	The NYSEO report recommended the West Valley site for interim storage.
1986	No provision for interim storage was included in the LLRW Management Act.
1988	An Interagency Task Force was created to investigate the possibility of interim storage after 12/31/92.
1989	The Interagency Task Force presented a preliminary report to the Advisory Committee. Opponents protested storage in Brooklyn. The Energy Authority released the NYS Plan for Storage and Management of Low-Level Radioactive Waste Generated in NYS after 1992.
1990	The Energy Authority was allocated funding by the legislature to conduct a storage study.

By 1983, the slow progress of compact negotiations had led some observers to conclude that New York State would not have a permanent disposal facility by the 1986 deadline specified in the LLRW Policy Act of 1980. Chapter 978 of the Energy Law enacted in August 1983, required NYSEO to study the issue of interim LLRW management.

The 1984 NYSEO draft report (NYSEO, 1984/a) analyzed various possibilities concerning interim management. It recommended negotiating with compacts and states with existing LLRW disposal facilities about possibilities for interim LLRW disposal and investigating the feasibility of providing interim storage for non-utility waste at nuclear power plant sites. The majority of the Advisory Committee, with a strong opposing minority opinion, favored the option of reopening the existing West Valley site for interim storage (NYSEO, 1984/b, Vol. 1., Attachment 1).

The final NYSEO report recommended on-site storage of utility waste by generators and seeking a license for a facility at West Valley to provide interim, above-ground, engineered, monitored storage of Class A, non-utility wastes. It also recommended pursuing arrangements for interim LLRW disposal with compacts and states where existing disposal facilities are located (NYSEO, 1984/b).

The 1986 Governor's Program Bill on LLRW Management contained the compromise recommendation that interim storage for non-utility waste should be made available at West Valley, provided this site was excluded from consideration as the site of the permanent disposal facility. In the closing days of 1985, however, the federal LLRW Policy Amendments Act was passed and signed into law in January 1986. The Amendments Act extended access to existing out-of-state sites until January 1993. It no longer appeared that interim storage would be necessary, and, afterwards, the designation of West Valley for interim storage was dropped from the proposed State legislation.

Due to delays in siting the permanent disposal facility, the problem of interim management emerged again in Autumn 1988. According to the LLRW Policy Amendments Act of 1985, by December 31, 1989, non-compact states must file applications to operate disposal facilities or certify to the NRC that they will provide for low-level waste storage or disposal within their borders after December 31, 1992. In case of noncompliance, New York State generators might lose access to the existing disposal facilities. In October 1988, an Interagency Task Force was established to examine how New York State could comply with the federal milestones and to investigate the possibility of storage after December 31, 1992. The Task Force included representatives from the Siting Commission, NYSDEC, NYSDOL, and NYSEO, and was chaired by the Energy Authority.

The Task Force focused primarily on the storage of non-utility waste, assuming that utilities have adequate space and tools for on-site storage. It considered a number of options,[36] which were examined in terms of multiple considerations.[37] Based on their analysis, the Task Force concluded that the only two tenable options were converting an existing (surplus) State facility to a LLRW storage facility or establishing a new State facility specifically for this purpose.

In its preliminary report, the Task Force concluded that the new facility option was more promising because the facility could be custom-designed for LLRW purposes to be expandable and readily decontaminated and decommissioned. The report recommended establishing a stand-alone facility to store non-utility LLRW on State-owned land, while utility waste should be stored on-site (Interagency Task Force, 1989/a).

---

[36] These included no action; arranging with out-of-State commercial disposal facilities; using out-of-State federal facilities; seeking Congressional relief from LLRW deadlines; encouraging a private storage development; using existing State LLRW facilities or modifying other State facilities for LLRW storage; using in-State federal facilities; and, developing a new State storage facility (Interagency Task Force, 1989/a).

[37] These included public health and safety; environment; federal deadlines; independence; expandability; transition to disposal facility; regulatory considerations; need for waste treatment; cost; and, likelihood of success.

In January 1989, the Task Force's preliminary report was presented to the Advisory Committee, which solicited comments from the public. The Advisory Committee expressed concerns about a stand-alone storage facility. The Committee proposed that power plant wastes not be excluded and that the facility be located at the same site as the permanent disposal facility (Interagency Task Force, 1989/b).

After receiving the Advisory Committee's recommendations, the Task Force reconvened to investigate the feasibility of implementing them. It concluded that a storage facility could not be established at the site of the permanent disposal facility until all regulatory approvals for the permanent facility had been obtained, which could not be before August 1994. The Task Force concluded that until storage becomes available, generators should maximize their on-site storage capabilities, and New York State brokers should provide supplemental storage capacity where it is needed (Interagency Task Force, 1989/b).

The Energy Authority spoke to two in-State LLRW brokers, Radiac and NDL, who indicated their potential willingness to store the waste of non-utility generators that could not expand their storage capacity. Both brokers also indicated they might be able to help generators who wanted to reduce their volumes through on-site compaction and by out-of-State processing, as well as to help generators deliver waste to out-of-State processors, who would return processed waste to generators or brokers for on-site storage (NYSLLWG, 1990).

In Summer 1989, several local citizens groups in the Williamsburg community of Brooklyn, where Radiac is located, increased their protest activities. In November 1989, the Assemblyman representing the district wrote to the Governor protesting the interim storage plan (Lentol, 1989). Generators also expressed concerns about the conclusions of the Task Force, arguing that many generators, particularly hospitals and bio-medical facilities, have limited storage capacity (NYSLLWG, 1989/b).

In preparation for the January 1, 1990, milestone under the Federal LLRW Policy Amendments Act and based on the recommendations of the Interagency Task Force, in November 1989, the Energy Authority distributed for public comment a proposed plan for storing, disposing of and managing LLRW after 1992 (NYSERDA, 1989/b). The key elements of the plan were that: the Energy Authority would establish permanent LLRW disposal faciliti(es) by October 1995 and would establish an interim LLRW storage facility at the permanent disposal site by August 1994; LLRW generators would minimize LLRW generation and maximize on-site storage capacity; and, the two in-State LLRW brokers would provide supplemental storage capacity for generators unable to store all their waste on-site.

In December 1989, the Governor filed a letter of certification accompanied with the above plan to NRC, DOE, and the states of South

Carolina, Nevada, and Washington (Cuomo, 1989/b-f). In January 1990, the Governor was advised by the relevant parties that New York State was found to be in compliance with the 1990 Federal milestone. In March 1990, however, the Governor received a letter (Griepentrog et al., 1990) from the representatives of Nevada, Washington, and South Carolina seeking information on a number of points in addition to the documentation that New York State had filed in compliance with the 1990 milestone established by the federal LLRW Policy Management Amendment Act of 1985.

The Energy Authority responded on behalf of the State to these concerns in April 1990, affirming that the State is committed to taking necessary actions to meet its lawful responsibilities for LLRW management, including providing the necessary funds to accomplish the specific tasks at hand (White, 1990/a). The letter described in detail the State's plans (White, 1990/a-c).

Meanwhile, there were public protests in the Williamsburg neighborhood of Brooklyn. A number of citizen and environmental groups around the State strongly recommended on-site storage of LLRW, arguing that nuclear power plants were already contaminated and, thus, represented a better site for waste storage than sites that were presently uncontaminated. Representatives of Cortland County also recommended examining the potential for storing institutional waste at the utilities (*Cortland County LLRW Newsletter*, July/August, 1991).

In light of the potential need for interim storage of LLRW in New York State, the 1990-91 State Operation Budget appropriated funds to the Energy Authority to study long-term storage options, including examining the existing capacity of LLRW generators to store waste on site, assessing their ability to expand capacity to accommodate storage for at least ten years, and investigating the economic viability of establishing a separate, centralized storage facility for Class A waste generated by hospitals, academic research, and medical research facilities (NYSERDA, 1990/c). A broad-based Study Review Panel and an independent Technical Review Committee would be established to provide input to the study design, statement of work, contractor solicitation, and critique study results (NYSERDA, 1990/c).

In October 1990, the Governor received a letter from the representatives of Nevada, Washington, and South Carolina expressing concerns about the adequacy of the New York State process to develop a permanent disposal facility (Stanley et al., 1990). The Energy Authority responded to these concerns in December 1990 indicating that, although the State is challenging the constitutionality of the Federal Low-Level Radioactive Waste Policy Act, New York is proceeding to implement an effective and timely LLRW management program in accordance with law (White, 1990/d). The response included an update (NYSERDA, 1990/c) which provided information on the status and progress of the LLRW Management Program. In their response in

January 1991, representatives of Nevada, Washington, and South Carolina determined that New York State was in compliance and assured that access to their disposal facilities was further provided (Griepentrog et al., 1991).

In June 1991, the Energy Authority selected the members of the Study Review Panel of the LLRW Storage Study. Members of the panel include officers from State and local governments, members of citizen and environmental groups, and waste generators. The Energy Authority contracted with a private contractor to facilitate Panel meetings and initiated steps to establish the independent Technical Review Committee to help to assure the scientific integrity of the study.

In June 1992, NYSDOH disseminated a notice to LLRW generators asking them to prepare plans for on-site waste management from January 1, 1993. Plans should include waste minimization procedures and on-site storage, and needs to remove administrative restrictions by licensing agencies (NYSDOH, 1992/b).

In July 1992, the Energy Authority released a Draft Interim Report on extended storage by New York State generators (NYSERDA, 1992/b). The report concluded that there are no absolute regulatory or technical obstacles to extended LLRW storage.

In November 1992, the Draft Design Basis Specification for a centralized storage facility for Class A LLRW from medical and academic generators was issued by the Energy Authority (NYSERDA, 1992/c). A facility could be designed to store wastes generated by medical and academic institutions for a 10-year period, with the possibility of being expanded for a 30-year period.

ANALYZING THE NEW YORK STATE CASE HISTORY FROM THE PERSPECTIVE OF THE COMPETING VALUES APPROACH

The New York State process was strongest from the empirical perspective, which emphasizes data-based processes and accountability in decisions. The Siting Commission and its contractors collected large amounts of data about geology, hydrology, and other characteristics of the investigated sites, as well as data about alternative permanent disposal technologies. The site- and method-selection processes developed by the Siting Commission involved the use of formal MAU models to identify optimal sites and methods.

The New York siting process was also well documented. New York assigned responsibility to State agencies for all aspects of the siting process, including site- and method-selection, public involvement, developing a compensation and incentive package, and interim management. As befits a state with a rich bureaucratic tradition like New York's, all the major decisions were generally well documented.

The strong emphasis on data collection, the use of formal decision-making models, and thorough documentation helped to make the process clear, increasing the accountability of the resulting decisions and recommendations. These characteristics also made it relatively easy for opponents and detractors of the process to criticize it, especially concerning the quality and appropriateness of that data. Much of the required input data for the models (e.g., relative importance of evaluation criteria) were necessarily subjective and uncertain, and several local governments and community groups criticized the agencies for relying on poor quality data and ignoring important information. Moreover, a number of detractors criticized the Siting Commission for not strictly following its own siting procedure, a conclusion supported by the 1992 GAO report.

According to the Competing Values framework, the goals and objectives of the political perspective -- which emphasizes adaptability in the decision making process -- compete directly with the goals and objectives of the empirical perspective. This means that processes that place a premium on accountability typically suffer in terms of their adaptability. This was precisely the case in New York State.

The New York State LLRW Management Act and the NYSDEC regulations prescribed the site- and method- selection processes in detail, and the Siting Commission developed detailed plans for site and method selection. Although this promoted accountability, the process was not very adaptable to unforeseen circumstances. For example, the process was not well suited for coping with the volunteer sites that unexpectedly materialized after the initial list of candidate sites was announced and had encountered significant local opposition.

The process was also not adaptable in responding to local interests and concerns. State law made no provisions for accommodating direct negotiations about compensation and incentives between the Siting Commission and local communities. This tied the hands of the Siting Commission in its efforts to deal with local concerns. After the siting process had stalled in the face of strong local opposition in Cortland and Allegany counties, a group of LLRW generators independently initiated negotiations with the Town of Ashford concerning a benefits package in exchange for siting the facility in West Valley. Negotiations between the generators and local governments afforded greater flexibility because they were not specifically constrained or circumscribed by statute. Although the various concerned parties appeared to be close to reaching an agreement, once again the inflexibility of the process contributed to gridlock. Before it could be implemented, any agreement between the generators and local governments needs to be enacted into law by the State Legislature, which to date has taken no action on the matter.

The New York State process also does not rate highly in terms of the rational perspective -- which emphasizes goal-centered processes intended to lead to efficient decisions -- primarily because the goals of State government concerning LLRW management changed significantly over time. For several years after passage of the 1980 federal LLRW Management Act, the State actively investigated the possibility of becoming a member of the proposed Northeast Compact. In 1986, after rejecting the idea of joining the Compact, the State passed its own LLRW Management Act, which contained a comprehensive LLRW management plan. The Siting Commission was established to oversee the site- and method-selection processes, and serious attempts were initiated to establish an in-state disposal facility for New York State generators. During this period the New York siting process was quite resolutely goal-focused.

Due to vigorous opposition in candidate areas, however, in 1990, site selection activities were suspended while still in preliminary stages of the process. Since then, New York State appears to have focused primarily on maintaining access to the existing commercial facilities by complying with the mandated federal milestones. The State also challenged the federal law in court, succeeding in overturning that part of the law that assigned responsibilities for LLRW management to the states.

Altering goals in response to changing conditions, as could be argued was what happened in New York, may be entirely reasonable behavior. Adhering to inappropriate goals in the face of radically changed circumstances is no virtue. Whatever the appropriateness of changing the focus of the siting process as happened in New York, it is difficult to argue that the process was efficient, at least as evaluated in terms of return on investment. During the period 1986 to 1992, over $40 million was spent, without a resolution to the problem of LLRW disposal.

Finally, the New York State process was not very successful from the consensual perspective -- which emphasizes participatory processes that are intended to help promote supportable decisions.

Various state agencies were responsible for assuring public involvement in the New York State siting process. NYSDOH conducted a statewide public information program; NYSDEC held public hearings concerning regulations and environmental impact assessment; and the Siting Commission held a number of hearings and open meetings throughout the process.

As a key vehicle for public participation, the 1986 State legislation established an LLRW Advisory Committee -- consisting of representatives of government agencies, waste generators, environmental organizations, and the general public -- the purpose of which was to advise government agencies on various aspects of the siting process. At least partially in response to pressure for greater public input to the process, the Advisory Committee was reconstituted in 1990 and renamed the Citizens' Advisory Committee. More citizen members were added and seats for State agency representatives were eliminated. In 1991, by which time the prospects for the timely siting of an in-State facility had dimmed and the likely need for some form of interim storage became ever more apparent, the Energy Authority initiated a LLRW Interim Storage Study, as directed by the State Legislature. To help to provide public input, the Energy Authority established a broad-based Study Review Panel.

All these various vehicles for public participation enjoyed limited authority. Neither the Advisory Committee, the Citizens' Advisory Committee, the Study Review Panel, nor any of the NYSDOH and NYSDEC initiatives granted decision making power, or even the power to make recommendations, to participants. Their power was limited to reviewing the decisions or recommendations of State agencies.

Similarly, the siting process did not give local governments any decision-making authority, even the power to decide whether they wanted to be considered as a potential site for the disposal facility. Over time, a number of local community groups and local governments did involve themselves in the siting process, of course, frequently as opponents and, in one case, as proponents, but such involvement was not by design.

In sum, a number of public participation initiatives were undertaken during the siting process in New York State, but, in all of these, the level of decision making power and authority vested in the public, either directly or through local government, was low. To date, public participation efforts do not appear to have been successful in promoting the development of a broad-based consensus in New York State regarding appropriate policies for LLRW disposal.

Chapter 4

CALIFORNIA

BACKGROUND

In 1979, approximately 4,300 cubic meters (83,300 curies) of LLRW was generated in California. By 1990, this figure changed to approximately 1,600 cubic meters and 5,500 curies (Table 12). The majority of the waste produced in 1990 came from industry (66 percent of the total volume and 90 percent of the total activity), whereas nuclear power reactors generated 24 percent of the total volume and four percent of the total activity (DOE, 1980, 1981, 1982, 1983, 1984/a, 1985, 1986/a, 1987, 1988, 1989, 1990, 1991; California, State of, DHS & BLM, 1991).

Table 12. TOTAL LOW-LEVEL WASTE GENERATED IN CALIFORNIA AND DISPOSED OF AT COMMERCIAL DISPOSAL SITES BY YEAR

	Volume (cubic meters)	Radioactivity (curies)	Volume Percentage by Source		
			Institutional and Industrial	Power Reactor	Government and Military
1979	4,342	83,281	100%	0%	0%
1980	6,129	32,002	75%	21%	4%
1981	6,051	3,271	63%	35%	2%
1982	5,407	18,070	75%	23%	2%
1983	3,781	41,518	74%	21%	5%
1984	4,489	13,677	71%	23%	6%
1985	7,105	53,614	60%	37%	3%
1986	3,228	39,214	55%	43%	2%
1987	2,813	29,461	58%	22%	20%
1988	2,355	25,614	53%	39%	8%
1989	3,340	9,949	69%	26%	5%
1990	1,585	5,538	66%	24%	10%

Source: DOE, 1980 - 1991.

A chronological summary of key events related to the LLRW disposal facility siting in California is given in Table 13.

Table 13. CHRONOLOGICAL SUMMARY OF KEY EVENTS IN CALIFORNIA

Date	Event
1982	California Legislature passed Assembly Bill 1513, which established plan for LLRW management.
1983	California Legislature passed Senate Bill 342, which required selection of a private company (license designee) to select a site and establish and operate a facility on State-owned land.
1984	The Department of Health Services (DHS) promulgated Title 17, Group 7 regulations for the license designee.
1985	US Ecology selected as license designee.
1986	US Ecology identified 18 desert basins as potential siting areas and initiated a Statewide public participation program to select a site.
1987	Based on the recommendations of the Site Selection Citizens' Advisory Committee and technical considerations, US Ecology selected the Ward Valley site. Legislation was enacted to create the Southwestern Interstate LLRW Compact, including California, Arizona, North Dakota, and South Dakota, with California as the host state.
1988	A Compact Commission with a majority of California members was formed to administer the Compact. Congress approved Southwest Interstate LLRW Compact.
1989	US Ecology filed a license application with DHS to construct a near-surface disposal facility with enhancements.
1990	DHS and the Bureau of Land Management published a joint EIR/EIS. Organized local opposition to the Ward Valley site started.
1991	DHS published the draft license. The City Council of Needles passed a resolution unequivocally opposing the site. Transfer of federal land by state indemnity selection was blocked by the State Land Commission.
1992	DHS agreed to hold additional adjudicatory hearings on the license application. US Ecology and two other generators filed a lawsuit against DHS and the Health and Welfare Agency. The Court granted a stay on the adjudicatory hearings. The Bureau of Land Management began the necessary process for selling the Ward Valley site to DHS.

OVERALL APPROACH

California responded to the federal LLRW Policy Act by passing Assembly Bill 1513 in 1982 (California, State of, Ch. 95, 1982). The law directed the

Department of Health Services (DHS) to develop an overall plan for LLRW management, treatment, and disposal; develop a classification scheme for the separation of LLRW; establish and operate LLRW interim storage facilities; appoint an advisory committee to advise DHS on LLRW management; and levy fees on California waste generators in order to pay for these activities.

DHS conducted a study in 1982 which recommended that: the search for a permanent burial site should begin immediately; California should make efforts to form a compact; and all efforts should be made to avoid the necessity of building and operating an interim storage facility. Arrangements should be sought with sited states to accept LLRW from California for a limited period (California, State of, DHS, 1982).

In 1983, the California State Legislature enacted Senate Bill 342 (California, State of, Ch. 1177, 1983). This bill was drafted by members of the California Radioactive Materials Management Forum (Cal Rad Forum), a Statewide organization of generators, in conjunction with legislative staff. It passed without controversy. This act required DHS to designate a private entity (a license designee) to select a site, and establish and operate a facility on State-owned land. The facility would be regulated by DHS. The law authorized DHS to require that a license designee post a bond of up to $1,000,000 to guarantee that it will carry out the activities needed to obtain a license.

DHS was authorized to set and collect fees from generators in California in order to support its activities associated with LLRW management, including the regulation, licensing, decontamination, decommissioning, and post-closure maintenance of the disposal facility, and the development and operation of interim storage facilities. The law also required the Governor to negotiate a compact agreement with other states and directed that compact agreements be ratified by the Legislature.

In 1984, a bill was introduced in the Senate to establish a compact between California and Arizona. The bill passed the Senate without any controversy, but was defeated in the Assembly. Opponents of the bill criticized it because, in spite of the fact that Arizona would generate about one quarter of the region's waste, there was no reciprocal arrangement in the legislation. In addition, opponents of the bill argued that the proposed legislation was ambiguous concerning finances (e.g., third-party liability) and measures to protect the environment (e.g., methods and procedure of transportation and waste packaging).

In 1986, a bill (Assembly Bill 1000) was introduced proposing a compact including California, Arizona, North Dakota, and South Dakota. The bill proposed that the largest generator (California) would be the site of the facility for the first 30 years. The second largest generator (most likely Arizona) would then host the facility for the next 30 years, and responsibility would subsequently alternate between these two. The amount of waste produced per year by the other states in the compact is shown in Table 14. In 1987, the Low-Level Radioactive Waste Disposal Compact Bill (Assembly Bill 1000) was ratified by the legislature of California (California, State of, Ch. 52, 1987). The law makes each party state responsible for developing and enforcing procedures for meeting the packaging and transportation requirements of the host state. The compact is authorized to collect surcharges from generators in the member states. Such funds are to be used to support the activities of the Commission, provide for insurance against third-party liability, and reimburse local governments for any costs or increased burdens incurred as a result of the facility.

Table 14. TOTAL LOW-LEVEL WASTE GENERATED IN THE NON-HOST STATES OF THE SOUTHWESTERN COMPACT AND DISPOSED OF AT COMMERCIAL DISPOSAL SITES BY YEAR

	Arizona		North Dakota		South Dakota	
	cubic meters	curies	cubic meters	curies	cubic meters	curies
1979	54	61	2	< 1	< 1	< 1
1980	49	225	4	< 1	< 1	< 1
1981	20	42	0	0	0	0
1982	3	1	0	0	0	0
1983	0	0	0	0	0	0
1984	16	38	0	0	0	0
1985	103	84	0	0	< 1	< 1
1986	136	108	0	0	< 1	< 1
1987	470	976	< 1	< 1	0	0
1988	803	829	0	0	0	0
1989	899	1,123	3	2	0	0
1990	786	247	1	< 1	33	< 1

Source: DOE, 1980 - 1991.

The law requires that a Compact Commission be formed, the primary responsibility of which will be to ensure that LLRW is safely disposed of in the region. Membership is to consist of one voting member from each state, appointed by the governors. The host county is also entitled to one voting member. The governor of California is entitled to appoint additional members to give California at least 51 percent of the Commission.

No other state may join the compact without the approval of at least two-thirds of the Compact Commission and ratification by the legislatures of all party states.

Soon after passage of the California law, similar laws were passed in Arizona, North Dakota, and South Dakota. Congress ratified the Southwestern Compact in November 1988 (Public Law 100-712, 1988). In 1991, the Compact Commission rejected requests from 15 states which had requested consideration to use the regional disposal facility. The Commission further voted to reject any future requests.

INSTITUTIONAL STRUCTURE

As an agreement state since 1962, California has the authority to license users of radioactive material. DHS is the agency responsible for issuance of licenses for the management of radioactive material.

Assembly Bill 1513 (California, State of, Ch. 95, 1982) also assigns responsibilities for oversight of LLRW disposal and interim storage to the DHS. The law requires DHS to develop an overall plan for the management, treatment, and disposal of LLRW. DHS was directed to define screening criteria for a permanent disposal site and to identify regions that meet these criteria. The law authorized DHS to establish and operate, or contract for, the establishment and operation of an interim storage facility.

The 1983 law (California, State of, Ch. 1177, 1983) assigned further responsibilities to DHS. The Department was required to develop emergency regulations for the licensing of LLRW disposal facilities. It was also instructed to establish regulations for soliciting, evaluating, ranking, and designating a private firm to locate, build, and operate a disposal facility.

In 1984, DHS promulgated regulations on Requirements for Land Disposal of Radioactive Waste (California, State of, Code of Regulations, Title 17, Group 7, 1984). The regulations established a procedure for selecting a license designee. Applicants were evaluated according to the following criteria: meeting the financial standards and qualifications; providing the best concept for site development and operation; presenting an effective program to deal with public concerns; and establishing a reasonable schedule of charges.

Financial requirements included posting a performance bond of $1,000,000 and an annual license fee of $250,000 (California, State of, Code of Regulations, Title 17, Group 7, 1984).

The costs of siting and development of a disposal facility will be invested by the license designee. Investment costs will be recaptured with accrued interest and profit from disposal fees over a 20-year amortization period (California, State of, Code of Regulations, Title 17, Group 7, 1984). Due to the monopoly status of the license designee, however, disposal fees will be subject to DHS rate review and approval (California, State of, Ch. 1177, 1983).

By July 1984, applications were received from Chem-Nuclear, Pacific Nuclear-Morrison Knudsen (as a joint venture), US Ecology, and Westinghouse. DHS reviewed the four applications and rated Westinghouse superior because it proposed an enhancement to shallow land burial, while the other three, which proposed shallow land burial without enhancements, were not rank ordered. All four firms were determined to be qualified, however.

In August 1984, Westinghouse declined designation. In October 1984, DHS reopened the license designee selection process. Chem-Nuclear filed a lawsuit against DHS to enjoin a second round, alleging that DHS failed to rank the applicants other than Westinghouse. In July 1985, the court ordered DHS to rank the remaining three applicants. In November 1985, DHS announced the following ranking: (1) Pacific Nuclear-Morrison Knudsen, (2) Chem-Nuclear, and (3) US Ecology. By December 1985, the Pacific Nuclear-Morrison Knudsen team dissolved, Chem-Nuclear withdrew for liability reasons, and US Ecology was selected license designee.

Assembly Bill 1513 (California, State of, Ch. 95, 1982) directed DHS to establish an expert committee to advise on LLRW management. The Low-Level Radioactive Waste Advisory Committee was established in 1983. Its members included representatives from the field of medicine and from research, industrial, environmental, and public health organizations. The 1987 law (California, State of, Ch. 52, 1987) directed that the senior environmental health officer of the host county be included on the Advisory Committee.

DESIGN OF THE SITE- AND METHOD-SELECTION PROCESSES

According to statute (California, State of, Ch. 1177, 1983), DHS was responsible for developing regulations for the site- and method-selection processes. The DHS adopted the NRC Part 61 regulations without any further specification of the processes (California, State of, Code of Regulations, Title 17, Group 7, 1984).

In 1983, the Cal Rad Forum conducted a public participation/education program three years before site selection started. The program included slide presentations on the Richland, Washington, and Beatty, Nevada, sites operated by US Ecology; sponsored visits to the Beatty site; distributed of a newsletter on LLRW management and a booklet published by the League of Women Voters; and made oral presentations (Cramer et al., 1989).

In 1986, US Ecology designed the site-selection process. It included the following phases: identifying candidate siting areas; identifying candidate sites; evaluating candidate sites; and site characterization.

After completing Statewide screening and narrowing the search to 18 candidate areas in Inyo, Riverside, and San Bernardino counties in 1986, US Ecology approached the League of Women Voters Southern California Task Force to provide advice on an effective public participation program. The Task Force, known for its citizen education and involvement activities, agreed to provide an impartial convenor/moderator, meeting arrangements, and related support services for the program (Schuiling & Anderson, 1991).

US Ecology and the convenor/moderator designed a public participation program which consisted of three components (Schuiling & Anderson, 1991). The first component included holding three sets of public meetings in the candidate areas. The first set of public meetings elicited public concerns about the facility. The second set of meetings elicited relative importance ratings, while at the third series of meetings, participants were requested to rank-order candidate sites.

The second component of the program was the formation of the Site Selection Citizens' Advisory Committee (SSCAC) to provide recommendations for US Ecology regarding site selection. Twelve citizens were selected to serve on the SSCAC. Two members were appointed by the Boards of Supervisors of each of the counties where candidate sites were located. The League of Women Voters selected three members, one from each of the affected counties. The other three were selected by the Sierra Club, the Native American Heritage Commission, and the Cal Rad Forum.

The third component focused on the involvement of elected officials, regulatory agencies and other government bodies, and the media on local, regional and state levels. Personnel from each of these groups were consulted regularly by US Ecology, and the SSCAC.

After announcing three candidate sites in February 1987, the League of Women Voters assisted in forming Local Advisory Committees (LACs) in the communities nearest to sites. Members of the LACs were nominated by local organizations and groups, and the meetings were facilitated by the same convenor who conducted the meetings of the SSCAC. The purpose of the LACs was to promote communication between US Ecology and local communities and to identify local information needs. After the announcement of the preferred site (Ward Valley) in March 1988, the LAC for this site continued to meet and make recommendations to US Ecology and DHS, while the SSCAC and the other LACs were dissolved.

The public participation program implemented by US Ecology sought direct public input and discussed all key decisions on site selection with the SSCAC and the LACs. The company's candidate site-selection process eventually ratified the public and SSCAC's consensus choices. As for decisions on other issues, including method of disposal and compensation, there was much less input from the public, primarily because these decisions were the responsibility of DHS and the Compact Commission, rather than US Ecology. The LACs, however, had some opportunity for input on disposal technology, and there was also some opportunity for local input on compensation through local representation on the Compact Commission.

SITE SELECTION FOR THE PERMANENT DISPOSAL FACILITY

The statute (California, State of, Ch. 95, 1982) prescribed that DHS conduct a preliminary screening based on a set of criteria.[38]   In December 1982, DHS developed criteria for preliminary screening, which expanded on the list prescribed by statute (California, State of, DHS, 1982).   With the exception of minor administrative procedures, the provisions were adopted from the NRC Part 61 regulations in the California regulations.   The complete list of screening criteria appear in Table 15.

In 1983, DHS initiated a Statewide screening that identified regions in the State with the potential to meet the siting criteria.   The study reviewed available documents and defined the southeastern part of the State (portions of Riverside, San Bernardino, Inyo, and Imperial Counties) as the regions containing the largest potential areas for further consideration (Anderson, 1988).

Based on DHS guidelines (California, State of, DHS, 1982), US Ecology conducted Statewide and regional screening studies between 1984 and 1986. Statewide screening resulted in the identification of three regions in which potentially suitable sites might be located: Eastern and Southeastern California, San Joaquin Valley, and Northeastern California.

The next phase of screening was based on criteria related to topography, geology, hydrology, soil, and conflicting land use (California, State of, DHS & BLM, 1991).

In June 1986, US Ecology narrowed the study to 18 desert basins located in Riverside, San Bernardino, and Inyo Counties.[39]

---

[38] These included use of land, water, and natural resources; proximity to major population centers; presence of earthquake faults; protection against surface and groundwater contamination; effectiveness of engineered barriers; transportation safety; and economic impacts.

[39] These included: Bristol Lake Basin, Broadwell Lake Basin, Cadiz Lake Basin, Chuckwalla-Ford Basin, Coyote Lake Basin, Cronese Lakes Basin, Danby Lake Basin, Mesquite Hills Basin, Pahrump Basin, Palen Dry Lake Basin, Panamint Valley, Salina Valley, Searles Lake Basin, Sheephole Basin, Silurian Basin, Silver Lake Basin, Soda Lake Basin, and Superior Lake Basin (California, State of, DHS & BLM, 1991).

Table 15. STATE OF CALIFORNIA LOW-LEVEL RADIOACTIVE WASTE SITING CRITERIA

1. Disposal facilities should be located away from large expanding population centers.
2. The waste disposal facility site shall be selected with consideration given to land use and resource development and not located where recovery of sub-surface minerals or groundwater resources could result in increased erosion or significant changes in the hydrogeological system. Areas containing critical habitats of endangered species or cultural resources should be avoided.
3. Economic impact analysis of proposed sites should be undertaken using acceptable procedures (such as those offered by the State of California, Office of Planning and Research). Use of compensatory mechanisms should be considered.
4. Disposal sites shall be accessible by all-weather roads.
5. The site shall demonstrate acceptable exposure limits (i.e., the performance objectives) in modeling simulations of facility performance as delineated in 10 CFR 61.41.
6. The site shall not be located over zones of active faulting where seismically induced phenomena would jeopardize the ability of the facility to meet performance objectives. Similarly, the site shall not be located near zones of igneous or volcanic activity that will jeopardize the ability of the facility to meet performance objectives.
7. The waste disposal facility shall not be located in a coastal high hazard area or wetland, 100-year flood plains, arroyos, dry washes, areas of poor drainage, within the upstream or downstream influence of man-made dams, or regulatory floodways, on the down gradient from mountain canyons where flash flooding is likely to occur, or in an area with conditions that will be conducive to flooding.
8. A suitable site must meet the following general hydrologic criteria:
    (a)   maximum fluctuations of the water table and associated capillary fringe should not intrude the proposed lower boundary of the disposal units (16m);
    (b)   the hydrologic setting of the proposed site should be fully characterized according to methods presented by NRC (1982);
    (c)   the disposal site should not be located in areas where it would degrade groundwater quality; and,
    (d)   excavated soil shall be evaluated for appropriateness as backfill and trench cap material.
9. Areas where wind or water erosion processes may present a significant potential to jeopardize facility performance should be avoided.
10. A semi-arid to arid climate is optimal for siting. The site should demonstrate a low frequency of extreme weather events during the short-term to ensure site operation integrity.
11. The natural attributes of the site shall allow full characterization, modeling, monitoring, and analysis of geologic, meteorologic, hydrologic, and radiologic factors.
12. The facility should be located where utilities can be made available.
13. The low-level waste disposal facility should not be located so remotely as to impair the administration of security and fire services.
14. The disposal site should be able to meet the performance objectives outlined in 10 CFR 61 throughout the projected 500 year hazardous interval of the waste. The selected site must be able to meet the volume requirements predicted for the facility service life.

Source: California, State of, DHS, 1982.

In 1986, US Ecology initiated a public participation program to assist in selecting a site. In June 1986, US Ecology organized five public meetings in the three candidate counties. At these meetings, information was provided about radioactive waste, waste disposal technologies, federal law requiring new disposal sites, and the California siting process. Concerns most frequently raised at the public meetings included the distance of the site from population centers; transportation and access to the site; groundwater quality and protection; presence of earthquake faults; economic and land use impacts; and adverse health effects (Anderson, 1987).

Next, a workshop to define site selection objectives and criteria was organized by US Ecology and its contractor, Environmental Science Associates (ESA), for the SSCAC. Members of SSCAC were informed about the concerns that had been most frequently raised at the public meetings. They were also briefed about the state and federal guidelines relating to site-selection criteria.

US Ecology and the SSCAC defined ten objectives (Anderson, 1987). Several criteria were suggested under each objective, as displayed in Table 16. SSCAC members then were asked to assign relative importance to the criteria. Relative importance was elicited qualitatively. Criteria were classified as either exclusionary or discretionary. Exclusionary criteria were based on legal or regulatory requirements and were not subject to change. Discretionary criteria were defined as avoidance criteria, constraints, or opportunities, and were rated as high, moderate, or low in importance. The SSCAC discussed the ratings and reached consensus on the importance of each criteria (Anderson, 1987).

In September/October 1986, five public meetings were organized where objectives and criteria developed by US Ecology and the SSCAC were presented to the public and relative importance ratings were elicited similar to those that had been elicited from SSCAC members. Ratings assigned to criteria by SSCAC and the general public are displayed in Table 17 (Anderson, 1987).

In the next phase, US Ecology and its contractors used exclusionary and high-avoidance discretionary criteria for screening the 18 topographically closed desert basins. The application of the above criteria eliminated two basins entirely -- the Saline Basin and the Sheephole Basin -- and large areas

Table 16. TABLE OF OBJECTIVES

OBJECTIVE 1: *AVOID PRESENT AND
FUTURE POPULATIONS*

Criterion

1.1   Exclude areas within the boundary of
       projected population growth.
1.2   Avoid individual residence or occupied
       structure.
1.3   Avoid established seasonally occupied
       structures or sites.
1.4   Seek availability of emergency
       response: local fire, police,
       ambulance service and health care
       services.

OBJECTIVE 2: *AVOID GEOLOGIC AND
                  FLOOD HAZARDS*

Criterion

2.1   Exclude areas within earthquake fault
       setback (1/2 mi. perpendicular; 2 miles
       from end of fault).
2.2   Exclude 100-year flood plain.
2.3   Exclude areas where upstream
       drainage would create problems.
2.4   Exclude volcanism centers.
2.5   Exclude landslide areas, sand dunes
       and unstable soil areas.
2.6   Avoid areas designated for emergency
       release of waters.

OBJECTIVE 3: *PROTECT SURFACE AND
                  GROUNDWATER QUALITY*

Criterion

3.1   Exclude areas where waste would
       contact groundwater.
3.2   Avoid areas with less than 150 ft. soil
       thickness.
3.3   Avoid areas with less than 100 feet to
       groundwater.

3.4   Seek sites with simple hydrologic
       systems which can be readily
       monitored and modeled.
3.5   Seek sites in areas of low intensity
       rainfall.
3.6   Avoid aqueducts and supply
       reservoirs.
3.7   Avoid groundwater recharge areas.
3.8   Avoid public water supply wells; avoid
       well fields or high-capacity production
       wells.

OBJECTIVE 4: *CONSIDER
                  TRANSPORTATION ROUTE AND
                  SAFETY FACTORS*

Criterion

4.1   Seek site close to divided highway.
4.2   Avoid roads with frequent hazardous
       conditions.
4.3   Avoid two-lane roads passing through
       heavily populated areas.
4.4   Avoid highways with high accident
       rates.
4.5   Avoid sites requiring bridge
       improvements.
4.6   Seek sites that require minor road
       improvements and/or construction.
4.7   Seek sites with available railroad
       access.

OBJECTIVE 5: *PROTECT AREAS OF
                  RESOURCE DEVELOPMENT*

Criterion

5.1   Avoid areas of existing cultivated
       agriculture or development areas.
5.2   Exclude areas of existing or potential
       economic mineral resource
       development (e.g., active mining).

OBJECTIVE 6: *CONSIDER AREAS OF SPECIAL OR PROTECTED LAND USE*

Criterion

6.1   Exclude BLM Class C, Areas of Critical Environmental Concern, East Mojave National Scenic Area.

6.2   Exclude National Monuments & State parks.

6.3   Exclude military reservations.

6.4   Exclude private lands held for preservation; UC field stations; Nature Conservancy.

6.5   Avoid BLM Wilderness Study Areas.

6.6   Seek BLM Class M lands.

6.7   Seek areas within designated utility corridors in CDCA.

6.8   Avoid areas proposed for wilderness under Cranston Bill.

6.9   Avoid sites close to the proposed state prison development.

6.10 Avoid important designated off-road vehicle use areas.

6.11 Avoid designated campgrounds.

6.12 Avoid popular rockhounding areas.

6.13 Avoid popular hiking areas.

6.14 Avoid popular hunting areas.

OBJECTIVE 7: *PROTECT ENVIRONMENTALLY SENSITIVE AREAS*

Criterion

7.1   Exclude "critical habitats" of federally listed endangered plant or animal species.

7.2   Avoid sensitive native plant communities.

7.3   Avoid important habitat features (water locations, migratory routes).

7.4   Avoid identified scientific study areas.

SOURCE: Anderson, 1987.

OBJECTIVE 8: *PROTECT CULTURAL AND HERITAGE VALUES*

Criterion

8.1   Exclude sites designated or recommended for the National Historic Register.

8.2   Avoid identified historic areas.

8.3   Avoid Native American resource areas, e.g., ritual, burial and other sacred areas, recognized by Native American communities.

8.4   Avoid Native American hunting and gathering areas.

8.5   Avoid sites adjacent to designated scenic highways.

8.6   Avoid identified BLM and National Park Service scenic resources.

OBJECTIVE 9: *SEEK COMPATIBILITY WITH COMMUNITY SOCIAL AND ECONOMIC GOALS*

Criterion

9.1   Seek compatibility with county general plans.

9.2   Seek sites near communities desiring the economic benefits of the facility.

9.3   Seek site close to existing available labor force.

OBJECTIVE 10: *PROTECT AIR QUALITY*

Criterion

10.1 Avoid sites located upwind of military installations (China Lake Naval Weapons Testing Center and Edwards Air Force Base) for which visibility is important to operations.

10.2 Avoid sites near Class I air quality management areas.

10.3 Avoid sites with frequent high surface winds.

Table 17. SITING OBJECTIVES AND CRITERIA SUMMARY

PUB: (Column 1) Summary of ratings from eight public meetings and one federal, state, and local agency meetings.   Total attendance - 480; worksheets turned in - 96.

CAC: (Column 2) Summary of Citizens Advisory Committee ratings.

Key: E = Exclusionary, H = High Importance; M = Moderate Importance; L = Low Importance; ? = No Majority

---

OBJECTIVE 1: *AVOID PRESENT AND FUTURE POPULATIONS*

Criterion

PUB	CAC	
E	E	Exclude areas within the boundary of projected population growth.
H	H	Avoid individual residence or occupied structure.
H	H/M	Avoid established seasonally occupied structures or sites.
E	E	Seek availability of emergency response: local fire, police, ambulance service and health care services.

OBJECTIVES 2: *AVOID GEOLOGIC AND FLOOD HAZARDS*

Criterion

PUB	CAC	
E	E	Exclude areas within earthquake fault setback (1/2 mi. perpendicular; 2 miles from end of fault).
E	E	Exclude 100-year flood plains.

PUB	CAC	
E	E	Exclude areas where upstream drainage would create problems.
E	E	Exclude volcanism centers.
E	E	Exclude landslide areas, sand dunes and unstable soil areas.
E	E	Avoid areas designated for emergency release of waters.

OBJECTIVE 3: *PROTECT SURFACE AND GROUNDWATER QUALITY*

Criterion

PUB	CAC	
E	E	Exclude areas where waste would contact groundwater.
H	H	Avoid areas with less than 150 ft soil thickness.
H	H	Avoid areas with less than 100 feet to groundwater.
H	H	Seek sites with simple hydrologic systems which can be readily monitored and modeled.
H	H	Seek sites in areas of low intensity rainfall.

PUB	CAC	
H	H	Avoid aqueducts and supply reservoirs
H	H	Avoid groundwater recharge areas.
H	H	Avoid public water supply wells: avoid well fields or high-capacity production wells.

OBJECTIVE 4: *CONSIDER TRANSPORTATION ROUTE AND SAFETY FACTORS*

Criterion

PUB	CAC	
$M/_H$	M	Seek site close to divided highway.
$H/_M$	$M/_H$	Avoid roads with frequent hazardous conditions.
$H/_M$	M	Avoid two-lane roads passing through heavily populated areas.
$H/_M$	$M/_H$	Avoid highways with high accident rates.
$M/_H$	M	Avoid sites requiring bridge improvements.
$L/_M$	L	Seek sites that require minor road improvements and/or construction.
$L/_M$	?	Seek sites with available railroad access.

OBJECTIVE 5: *PROTECT AREAS OF RESOURCE DEVELOPMENT*

Criterion

PUB	CAC	
H	H	Avoid areas of existing cultivated agriculture or development areas.
E	E	Exclude areas of existing or potential economic mineral resource

development (e.g., active mining).

OBJECTIVE 6: *CONSIDER AREAS OF SPECIAL OR PROTECTED LAND USE*

Criterion

PUB	CAC	
E	E	Exclude BLM Class C, Areas of Critical Environmental Concern, East Mojave National Scenic Area.
E	E	Exclude National Monuments & State parks.
E	E	Exclude military reservations.
E	E	Exclude private lands held for preservation; UC field stations; Nature Conservancy.
H	H	Avoid BLM Wilderness Study Areas.
M	M	Seek BLM Class M lands.
M	M	Seek areas within designated utility corridors in CDCA.
?	?	Avoid areas proposed for wilderness under Cranston Bill.
$I/_M$	?	Avoid sites close to the proposed state prison development.
?	$M/_L$	Avoid important designated off-road vehicle use areas.
$H/_M$	$M/_H$	Avoid designated campgrounds.
$H/_M$	M	Avoid popular rockhounding areas.
$H/_M$	M	Avoid popular hiking areas.

PUB   CAC

$H/_M$ M   Avoid popular hunting areas.

OBJECTIVE 7: *PROTECT ENVIRONMENTALLY SENSITIVE AREAS*

Criterion

E   E   Exclude "critical habitats" of federally listed endangered plant or animal species.

$H/_M$ $M/_H$ Avoid sensitive native plant communities.

H $M/_H$ Avoid important habitat features (water locations, migratory routes).

$H/_M$ M   Avoid identified scientific study areas.

OBJECTIVE 8: *PROTECT CULTURAL AND HERITAGE VALUES*

Criterion

E   E   Exclude sites designated or recommended for the National Historic Register.

$H/_M$ $H/_M$ Avoid identified historic areas.

$H/_M$ $H/_M$ Avoid Native American resource areas, e.g., ritual, burial and other sacred areas, recognized by Native American communities.

$H/_M$ M   Avoid Native American hunting and gathering areas.

PUB   CAC

$M/_H$?   Avoid sites adjacent to designated scenic highways.

$H/_M$?   Avoid identified BLM and National Park Service scenic resources.

OBJECTIVE 9: *SEEK COMPATIBILITY WITH COMMUNITY SOCIAL AND ECONOMIC GOALS*

Criterion

$H/_M$ $H/_M$ Seek compatibility with county general plans.

$H/_M$ $H/_M$ Seek sites near communities desiring the economic benefits of the facility.

M   M   Seek site close to existing available labor force.

OBJECTIVE 10: *PROTECT AIR QUALITY*

Criterion

$H/_M$ $M/_L$ Avoid sites located upwind of military installations (China Lake Naval Weapons Testing Center and Edwards Air Force Base) for which visibility is important to operations.

$H/_M$ $M/_H$ Avoid sites near Class I air quality management areas.

$H/_M$ $M/_H$ Avoid sites with frequent high surface winds.

Source: Anderson, 1988.

within the remaining 16 basins. Mapped overlays were used to identify 16 candidate siting areas which were believed to contain technically suitable, licensable areas (California, State of, DHS & BLM, 1991).[40]

In the next set of ten public meetings, organized in December 1986/January 1987, citizens were asked to complete questionnaires that identified the most and least favorable candidate siting areas, and to state their reasons. The public favored the Panamint Valley site which was relatively far from each community where public meetings took place.

After reviewing the results of the public workshops, SSCAC members were asked to divide the candidate siting areas into three categories: the top five areas; the middle six areas; and the bottom five. As a basis for discussions, both groups were asked to give the reasons for their ratings. There was consensus in the SSCAC with regard to the top sites. The SSCAC favored the Ward Valley, Silurian Valley, Danby, and Fenner Valley sites; Ward Valley and Silurian Valley received the highest ratings (Anderson, 1987).

Based on the SSCAC's and the public's ratings, in February 1987, US Ecology identified three candidate sites in Ward Valley and Silurian Valley (San Bernardino County) and the Panamint Valley (Inyo County). Candidate siting areas in Riverside County were eliminated.

During this phase of the siting process, there was public protest in several candidate siting areas. Petitions and letters of protest were received, particularly from Yucca Valley, Joshua Tree, and Twentynine Palms. However, there was no significant opposition in the communities around the three candidate sites.

After announcing the candidate sites, the League assisted in forming Local Advisory Committees (LACs) in the communities nearest to sites, namely, in Needles (Ward Valley), Baker (Silurian Valley), and Trona (Panamint Valley). LACs discussed the characteristics of the potential sites, disposal technologies, and the perceived advantages and disadvantages of having a site near their community.

By October 1987, US Ecology completed the characterizations of the three sites and provided SSCAC with detailed information about the site-selection criteria for each site. SSCAC members were asked to select a

---

[40] These included Bristol (Western part of Bristol Lake Basin), Broadwell (Central Broadwell Lake Basin), Cadiz (Southwestern Cadiz Lake Basin), Chuckwalla-Ford (Southern part of Chuckwalla-Ford Basin), Coyote-Alvord (Northwestern part of the Coyote Lake Basin), Danby (Southeastern Danby Dry Lake Basin), Fenner Valley (Eastern part of Bristol Lake Basin), Pathrump (Southwestern Pathrump Valley Basin), Panamint (West-Central Panamint Valley Basin), Central Searles (South-Central Searles Lake Basin, South Searles (Southern Searles Lake Basin), Silurian (Eastern Silurian Dry Lake Basin), Silver (Northwestern Silver Lake Basin), Soda North (Northern Soda Dry Lake Basin), Superior (Northwestern Superior Lake Basin), and Ward Valley (North-Central Danby Dry Lake Basin).

preferred site and a backup site. Both the Ward and the Silurian Valley sites received five votes, while the Panamint Valley site was dropped from further consideration (Anderson, 1988).

There were differences of opinion in the Committee about the suitability of the Ward Valley site. Some felt that the facility would create a danger to the habitat of the desert tortoise, a threatened species. The Native American member of the Committee objected to all candidate sites because they are considered tribal lands by Native Americans.

Between October 1987 and March 1988, US Ecology carried out more detailed characterizations of the two remaining sites. In March 1988, it designated the Ward Valley site as the preferred location and Silurian Valley as an alternative site. Both sites are located on federal land.

After studying the Ward Valley and Silurian sites in detail, in August 1989, US Ecology developed the facility design along with a Proponent's Environmental Impact Assessment and mitigation program. DHS formed a special committee to provide advice on project planning and on an appropriate research design for assessing impacts on the desert tortoise. The special committee's primary recommendations, including fencing the nearby freeway to protect tortoises, surrounding the disposal and support area by a chain-link fence, and relocating tortoises during the construction phase, were included in US Ecology's plans.

In September 1989, US Ecology filed a license application with DHS to construct the facility at the Ward Valley site. Under the California Environmental Quality Act, as licensing agency, DHS is required to prepare an Environmental Impact Report (EIR). Since the project involved a transfer of federally-owned land to the state, an Environmental Impact Statement (EIS), pursuant to the National Environmental Policy Act (NEPA, Section 15090), was completed by the Bureau of Land Management California Desert District (BLM). DHS and BLM agreed to prepare a joint Environmental Impact Report/Environmental Impact Statement (EIR/EIS). The Draft EIR/EIS was published in June 1990 (California, State of, DHS & BLM, 1990) and included detailed descriptions of the purpose and need for the facility; overview of the proposed project at the Ward Valley site; overview of the alternative Silurian Valley site; and environmental impacts and mitigation measures. In July 1990, DHS conducted public hearings and solicited public comments on the Draft EIR/EIS.

The potential for contamination of the Colorado River and the groundwater; flood protection; and impacts on the desert tortoise habitat were the major concerns associated with the selection of the Ward Valley site (California, State of, DHS & BLM, 1991). Additional concerns were raised about the accuracy of estimates of the waste stream and about the appropriateness of the proposed environmental monitoring system (California, State of, DHS & BLM, 1991).

Until this time, there was no significant public opposition to the Ward Valley site. In June 1990, however, a local group, People Against Radioactive Dumping (PARD), started to protest the planned facility. The group initiated a community-wide petition drive in opposition to the project and spoke out at a meeting of the LAC and numerous meetings of the Needles City Council. Their concerns included fears that a large portion of waste would come from decommissioned nuclear power plants and out-of-compact generators; that the facility would contaminate the Colorado River and the groundwater; that airborne radiation from the site would cause severe health problems to children; and that liability would become the responsibility of California and Arizona taxpayers (*The Sun*, June 23, 1990).

For several months, the Needles City Council meetings focused primarily on the LLRW disposal facility siting issue. In February 1991, the Council voted three to two not to support the siting in Ward Valley.

In April 1991, DHS published the draft license (California, State of, DHS, 1991). The Final EIR/EIS was also published in April 1991 by DHS and BLM (California, State of, DHS & BLM, 1991). The Final EIR/EIS addressed concerns raised by the public during the comment period on the Draft EIR/EIS. Additional measures were incorporated into the project design to monitor the facility performance and to protect the desert tortoise. In July 1991, public hearings were conducted. Public comments were elicited on both documents by DHS through August 1991.

As a formal response to the EIR/EIS, in June 1991, the Needles City Council sent a letter to DHS in which the City asked for a surcharge on LLRW disposal to compensate the city for adverse economic impacts and requested funding to be used to promote tourism and economic development (City of Needles, 1991). In September 1991, the City of Needles filed a follow-up supplement to their previous request including comments to the EIR/EIS and US Ecology's license application (Tanaka, 1991). The letter expressed concerns about potential water contamination, socioeconomic impacts of the project on the City, impact of the facility on the desert tortoise, and the questionable operating history of US Ecology.

The Needles City Council did not receive a positive response from DHS. US Ecology urged negotiations on additional local assistance between the City of Needles and the Cal Rad Forum. Negotiations with the City, however, never started. On November 18, the City Council of Needles passed a resolution unequivocally opposing the site (Council of the City of Needles, 1991). Cal Rad Forum entered into discussions with the School District and local businessmen to develop a compensation plan.

Another controversy that has involved participants from the broader political community within the State has related to liability issues. There is some concern that, as the host state, California would be ultimately responsible for expenses associated with future accidents.

In October, the Assembly Natural Resources Committee held a formal hearing on the Ward Valley project. Following the meeting, several legislators requested DHS to suspend further actions until the State's liability was reviewed (Woodruff et al., 1991).

In a letter to Governor Wilson, in November 1991, 20 California congressmembers expressed their concerns about the unresolved liability issues. They expressed their support for proposals requiring US Ecology to assume liability for any and all damages that might result from the facility and requiring the generators to finance reserve funds sufficient to pay for the cost of cleaning up any potential problem. They urged the Governor to intervene in the ongoing debate (Miller et al., 1991).

Citizen and environmental groups opposed to the facility fought for adjudicatory hearings on the license application to address additional health and safety issues. In December 1991, a Sacramento Superior Court judge ruled that the original three hearings were sufficient, and additional hearings were not required by state law. But the Senate Rules Committee successfully negotiated an agreement requiring DHS to hold the hearings, after threatening not to confirm the acting secretary of the Health and Welfare Agency and the acting DHS director (*The Radioactive Exchange*, Vol. 11, No. 15, 1992).

In July 1992, US Ecology and two nuclear medicine groups filed a lawsuit against the director of DHS and the secretary of the Health and Welfare Agency as well as against the two agencies, charging that their consent was "illegally coerced." In September 1992, the Third District Court of Appeals for the District of Sacramento granted a stay on the hearings (*The Radioactive Exchange*, Vol. 11, Nos. 15 & 16, 1992). As of December 1992, the lawsuit was in progress.

In 1991-92, the transfer of federal land became controversial. The initial final EIS identified state indemnity selection as the method of land transfer. Under that process, the State Land Commission would have acquired ownership of the land and then transferred it to the state. This became an unsurmountable problem when in early 1992 two of the three commissioners voted against the facility. In August 1992, the U.S. Bureau of Land Management began a process by which DHS can directly purchase the Ward Valley site (*The Radioactive Exchange*, Vol. 11, No. 15, 1992).

PERMANENT DISPOSAL METHOD SELECTION

Selection of a disposal technology was the responsibility of DHS. In its bid in 1984, US Ecology recommended near surface burial. Shallow land burial, as implemented at the Beatty site, was considered as the reference method of disposal during the site-selection process. In 1987, because of concerns raised by the public about the ability of the site to protect public health and the environment, DHS hired Ebasco Services, Inc., to analyze alternatives and enhancements to shallow land burial.

Ebasco (1988) evaluated three alternatives to shallow land burial: above-ground vaults; below-ground vaults; and earth-mounded concrete bunker. Ebasco concluded that all three methods were inappropriate for a desert environment because of the deterioration of concrete structures, the need for a longer institutional control period, higher costs, shorter routes for radionuclides to escape to the environment, and limited practical experience with application of the technologies (California, State of, DHS & BLM, 1991).

Ebasco (1988) studied the effects of a series of enhancements of shallow land burial on the most frequently mentioned accident scenarios.[41] A cost-benefit analysis was performed weighing the costs for development and operation of the enhancements against the benefits of reducing the estimated radiological doses associated with each accident scenario. The analysis concluded that deeper depth disposal, improved covers, and use of auger holes are promising enhancements in terms of the costs and benefits associated with dose reduction. Based on this analysis, DHS directed US Ecology to incorporate enhancements into the facility design (California, State of, DHS & BLM, 1991).[42] Enhanced shallow land burial is, thus, the designated disposal method in California.

US Ecology presented the proposed facility design in the Proponent's Environmental Assessment and the license application. After the Draft EIR/EIS (California, State of, DHS & BLM, 1990) was issued in June 1990, several concerns were raised about the proposed design. Most of these concerns were associated with the possible need for additional structures to protect groundwater from the migration of radionuclides and the advisability of expanding the environmental monitoring system.

The California Water Quality Control Board and the regional Colorado River Water Quality Control Board, as well as the U.S. Environmental Protection Agency, among others, criticized the proposed enhancements to

---

[41] The enhancements included: small trenches; trench with sand backfill; trench lined with clay; trench with concrete backfill; deeper depth burial; concrete overpack; trench with improved cover; stability requirement for Class A waste; separate trench for Classes B and C waste; unlined auger holes for waste more than 30 Rem/hour; and lined auger holes for waste more than 30 Rem/hour. The accident scenarios included: groundwater contamination; accidental release of radioactivity to the atmosphere; routine worker radiation exposure; effects of earthquakes on buried wastes; intruders trespassing on the site; plants absorbing radionuclides; animals entering the waste; and exposure to radiation resulting from wind and water erosion and flooding.

[42] These include: deeper depth burial, with a minimum cover of 5 meters of native soil over the waste; separate trench provided for Classes B and C waste and packages with more than 30 Rem/hour; improved engineered cover for Classes B and C waste; small trenches; and soil surcharge placed on the filled portion of the trenches.

shallow land burial and the procedures that would be used to monitor the movement of radionuclides through the trenches and the vadose zone or the soil between the ground surface and the water-bearing strata (California, State of, DHS & BLM, 1991). In response to these criticisms, DHS staff evaluated using a double liner with an internal leachate collection system as an additional enhancement. DHS rejected this approach because it would be expensive (estimated $25 million) and would create water accumulation and related environmental problems (California, State of, DHS & BLM, 1991).

To resolve the issue of monitoring, DHS established a Special Committee on Vadose Zone Monitoring, including experts from the U.S. Environmental Protection Agency, U.S. Nuclear Regulatory Commission, Sierra Club, Cal Rad Forum, and various industrial and academic institutions. The Committee investigated various alternatives and enhancements to the proposed system. Based on the findings of the Committee, DHS issued licensing guidance to US Ecology that monitoring be provided for all trench covers, for areas adjacent to the trenches, and below all trenches (California, State of, DHS & BLM, 1991). In the draft license, DHS indicated that "All potentially significant pathways, including direct radiation, air, vegetation, animals, ...soil and water (including the unsaturated zone) shall be monitored" (California, State of, DHS, 1991; p. 26).

As of late 1992, discussions about the safety and suitability of the method continue.

COMPENSATION AND INCENTIVES

The statute did not provide for compensating host communities; however, it mentioned that "a local agency or school district may pursue any remedies to obtain reimbursement available to it" (California, State of, Ch. 1117, Sec. 12, 1983).

The LLRW Disposal Compact Act (California, State of, Ch. 52, 1987) directed the Compact Commission to use disposal fee surcharges to establish a local government reimbursement fund. The fund could be used for reimbursing local governments for expenses related to hosting the regional facility including, but not limited to, the maintenance and improvement of roads and bridges, fire protection, law enforcement, monitoring by local health officials, and emergency preparation and response.

As the regulator, DHS interpreted compensation only to involve reimbursing the local community for measurable impacts. Providing benefits to the local community was conceptualized as environmental impact mitigation rather than as incentives. For example, because of the possible impact on local emergency services, the EIR/EIS included the mitigation requirements that the license must provide free training and buy equipment for all the local emergency response personnel (California, State of, DHS & BLM, 1991).

The issue of additional compensation came into the forefront in 1991 when, as a result of a local controversy on the siting of the disposal facility in Ward Valley, the Needles City Council claimed a right to additional reimbursement for negative perceptions and socioeconomic impacts not mentioned in the EIR/EIS. In its letters to DHS, the City Council described damages done to the local economy (e.g., tourism and housing) by the negative perceptions and stigma created by the controversy to date. The Council asked for a $10 per cubic foot surcharge on LLRW disposal and funding for a full-time employee to promote tourism and economic development (City of Needles, 1991; Tanaka, 1991).

No formal response was given to the City. US Ecology supported the position of the City that the language of the Compact Law did not exclude providing compensation to the local community for impacts resulting from negative perceptions. US Ecology urged Cal Rad Forum to negotiate compensation with the City. The generators acknowledged that some benefits should be provided and entered into negotiations with the School District and local businessmen. However, due to lack of interest on behalf of elected officials, negotiations between the host community and the generators stopped.

INTERIM WASTE MANAGEMENT

Interim management became an issue in California in 1982, when it became apparent that the State of California could not meet the deadlines created by the Federal LLRW Policy Act of 1980 and that interim storage would, therefore, be necessary. The statute of 1982 (California, State of, Ch. 95, 1982) required DHS to develop contingency plans for short-term storage of the waste generated in California. The law authorized DHS to establish and operate LLRW interim storage facilities and prescribed several siting criteria.[43]

In December 1982, DHS investigated the costs and hazards of establishing an interim storage facility (California, State of, DHS, 1982). The study found that building a facility for interim storage and then moving the waste from the interim facility to the permanent disposal site would duplicate much of the efforts and hazards. The study, therefore, concluded that all efforts should be made to avoid the necessity of building and operating an interim storage facility.

---

[43] These included: proximity to population; geologic stability; proximity to ground and surface water; availability of transportation; and general public health and economic considerations.

After the Federal LLRW Policy Act Amendments extended the deadline for permanent disposal until January 1, 1993, no further actions were taken on interim management until 1991. By 1991, however, it became clear that operation of the disposal facility by January 1, 1993, was unrealistic. In 1991, most generators developed contingency plans for on-site storage after January 1, 1993.

ANALYZING THE CALIFORNIA CASE HISTORY FROM THE PERSPECTIVE OF THE COMPETING VALUES APPROACH

During its early stages, when the California siting process was primarily overseen by a private contractor, the process was strong from the rational perspective, which emphasizes goal-centeredness and efficiency. In 1985, US Ecology was selected as license designee and, from that time on, had to invest its own funds in the siting process. Acting prudently in its own financial interests, US Ecology conducted a site-selection process that was highly goal-centered and that progressed at a rapid pace. By 1989, a site was selected, the disposal technology had been chosen, and US Ecology had submitted a license application. Overall, US Ecology spent approximately $25 million on siting process activities.

After the process entered the licensing phase, however, when control passed from the license designee to a number of different State agencies, it became less goal-focused and more susceptible to a variety of political interests and concerns. Since 1991, when local opposition to the chosen site at Ward Valley became evident, eventually leading to Statewide controversy, the process has slowed considerably.

With respect to the empirical perspective, the process placed heavy emphasis on the collection and analysis of data. Major responsibilities were divided among the license designee and various State agencies. US Ecology was responsible for site selection and constructing and operating the facility. DHS and BLM were responsible for method selection, land transfer, regulation, and licensing. Both the license designee and the State agencies drew on a large amount of technical information in preparing comprehensive environmental assessment studies, although decisions made by the State agencies were generally better documented than were those of US Ecology.

Although accountability was emphasized, this was not defined to mean that the process had to identify technically optimal sites and methods. Site- and method- selection processes were aimed instead at finding satisfactory, licensable sites and methods. Potential sites were evaluated in terms of multiple criteria, but US Ecology did not use formal MAU models. Similarly, the State conducted cost-benefit analyses to evaluate the effectiveness of various enhancements to shallow-land burial -- the chosen method -- but additional non-financial and non-technical considerations (e.g., public acceptability) also influenced decisions about method selection.

During the licensing process, several critics raised concerns about the accuracy of some data (e.g., nature and size of the waste stream) and the safety of the facility. In response, DHS established a special committee that included experts from federal agencies, environmental organizations, and various industrial and academic institutions to investigate potential enhancements to the monitoring system. The agency also agreed to hold an adjudicatory hearing on the license application and to address some additional health and safety concerns.

The California process receives mixed evaluations from the consensual perspective, which emphasizes participatory processes intended to lead toward decisions with broad-based support. California law explicitly required only public comment and hearings, but to promote support for the site-selection process, US Ecology organized a highly participatory process. A series of public meetings were held in candidate areas to elicit input from the general public. Elected officials, regulatory agencies, and other stakeholders were consulted regularly. A Site Selection Citizens' Advisory Committee was formed that included representatives of citizen and environmental groups, Native American organizations, generators, and the potentially affected counties. The Committee had the power to provide recommendations about site selection to US Ecology, and the company ratified the Committee's recommendations. Local Advisory Committees were also formed in the communities nearest to candidate sites, although these had little or no opportunity to influence decisions on disposal technology or to negotiate compensation and incentives.

Substantial levels of Statewide public participation seemed to contribute to widespread support for the US Ecology decision selecting the Ward Valley site. Later, lack of local community involvement in subsequent decisions after site selection appeared to lead to an erosion of public support for this siting decision, however.

The public also exercised considerable influence on the method selection process. Public concerns about the ability of the facility to protect public health and the environment led to a study which eventually resulted in a number of enhancements to shallow land burial, which is the designated disposal method in California.

Finally, the California process was designed to be highly flexible and adaptable, thus rating highly from the political perspective. The State provided few prescriptions concerning the site- and method-selection processes. US Ecology conducted a highly adaptable site-selection process, and the method-selection process conducted by DHS was also comparatively rather flexible. Perhaps partially as a result, decisions about site and method initially appeared to enjoy a substantial sense of legitimacy across the State. After the license application was submitted in 1989, however, when the various proponents failed to behave responsively to various local concerns or to negotiate compensation and incentives with the local community, the legitimacy of the process began to be questioned and undermined, first by local government and, later, by the broader political community.

Chapter 5

ILLINOIS

BACKGROUND

In 1979, approximately 6,800 cubic meters (9,000 curies) of LLRW was shipped from Illinois for disposal. The approximate quantities shipped in 1990 were 2,800 cubic meters and 8,000 curies (Table 18).

Table 18. TOTAL LOW-LEVEL WASTE GENERATED IN ILLINOIS AND DISPOSED OF AT COMMERCIAL DISPOSAL SITES BY YEAR

	Volume (cubic meters)	Radioactivity (curies)	Volume Percentage by Source		
			Institutional and Industrial	Power Reactor	Government and Military
1979	6,758	9,044	47%	36%	17%
1980	8,140	11,326	30%	70%	0%
1981	7,595	34,649	40%	60%	0%
1982	5,068	196,564	31%	69%	0%
1983	6,196	13,773	8%	91%	1%
1984	6,436	48,937	27%	69%	4%
1985	10,205	126,445	38%	57%	5%
1986	6,433	42,888	25%	75%	0%
1987	5,387	33,917	26%	74%	0%
1988	3,179	8,936	10%	89%	1%
1989	3,817	147,115	12%	86%	2%
1990	2,785	8,253	8%	91%	1%

Source: DOE, 1980 - 1991.

Illinois has more operating nuclear power plants than any other state in the United States. In 1990, 12 commercial power reactors were operating. Nuclear power reactors provided about 91 percent of the volume of Illinois' LLRW in 1990 (DOE, 1980, 1981, 1982, 1983, 1984/a, 1985, 1986/a, 1987, 1988, 1989, 1990, 1991).

A chronological summary of key events related to the LLRW disposal
facility siting in Illinois is given in Table 19.

Table 19. CHRONOLOGICAL SUMMARY OF KEY EVENTS IN
ILLINOIS

Date	Event
1980	Governor of Illinois established Illinois Department of Nuclear Safety (IDNS). Illinois participated in negotiations to form Midwest Interstate LLRW Compact.
1983	Illinois Legislature passed the Illinois LLRW Management Act. Illinois opened discussions with Kentucky to form a two-state compact.
1984	Legislation was enacted creating the Central Midwest Interstate LLRW Compact, including Illinois and Kentucky, with Illinois as the host state. A three member Commission was created to administer the Compact.
1985	Congress approved Central Midwest Interstate LLRW Compact. The Illinois LLRW Management Act was amended to prohibit the use of shallow land burial as a disposal method; require selection and characterization of multiple alternative sites; and make grants to local governments for independent review of site suitability.
1986	The IDNS issued Title 32, Part 606 regulations on the design, construction and operation of a disposal facility.
1987	IDNS hired Battelle Memorial Institute to perform studies for site selection.
1988	Westinghouse was selected as developer of LLRW disposal facility. Two sites (Martinsville and Geff) were selected for characterization.
1989	Westinghouse withdrew from contract; Chem-Nuclear Systems was named developer/operator. IGW Surveys publicly withdrew support for IDNS draft report summarizing results of characterization studies. The Governor asked a former Illinois Supreme Court Justice to hold public hearings and determine whether proposed sites were safe. Illinois Senate passed a resolution to investigate the LLRW program, and IDNS was ordered to stop all selection activities until the investigation was completed.
1990	The legislature amended Illinois LLRW Management Act and established an independent LLRW Disposal Facility Siting Commission to evaluate the safety and suitability of any site proposed by IDNS.
1991	IDNS recommended the Martinsville site to the Siting Commission. Hearings on Martinsville site initiated by Siting Commission. Chem-Nuclear submitted license application for a design including above-ground vaults containing modular concrete canisters, with multi-layered vault covers. IDNS conducted an interim storage survey with the generators of Illinois and Kentucky.
1992	City of Martinsville unanimously approved a "resolution of support" and a community agreement with IDNS. Siting Commission expressed serious concerns about the scientific integrity of the project and unanimously rejected the Martinsville site.

OVERALL APPROACH

Following the passage of the 1980 federal LLRW Policy Act, Illinois participated in negotiations that eventually led to the establishment of the Midwest Interstate LLRW Compact. As many as 17 states expressed some interest in the Midwest Compact. As the largest generator among the potential compact members, Illinois was most likely to host the facility if it decided to join (Illinois, State of, Department of Nuclear Safety [IDNS], 1991).

In 1983, the Illinois Legislature passed the Illinois Low-Level Radioactive Waste Management Act (Illinois, State of, Public Act [P.A.] 83-991, 1983). The law did not enter Illinois into any compact, but gave the IDNS the responsibility to develop a comprehensive program for storing, treating, transporting, and disposing of LLRW, which would include: developing a generator registration system; promulgating regulations for waste treatment, transportation, storage, and disposal; developing a plan for interim management; selecting a site for, planning, constructing, and operating a permanent disposal facility; designing procedures for compensating the affected communities for siting; and collecting fees from generators in order to cover the costs of LLRW management.

The act established two special funds in the State Treasury both to be supported by generator fees. The first is the LLRW Facility Development and Operation Fund which supports the work of IDNS and its contractors to establish a disposal facility. The second is the Low-Level Radioactive Waste Facility Closure, Post-Closure Care and Compensation Fund, which is a long-term fund to handle contingencies that might arise, such as compensating third parties for damages due to a radioactive material release from the facility.

The law authorized the Governor to enter Illinois into agreement state status with the federal government for regulating and licensing LLRW facilities. It prohibited IDNS from licensing a facility located within 1-1/2 miles of a municipality unless the siting is approved by the municipality.

The law stated that joining the compact would be subject to approval by the Legislature and a five-member advisory committee including representatives of both houses of Legislature and a designee of the Governor. If Illinois entered a compact and was selected as the host state, the law stated that it could withdraw unless at least one other state was designated to host a facility in the future.

There was substantial public concern, particularly from the environmental community, about some provisions of the proposed Midwest Compact. There were three main concerns: shared liability, limited duration of obligation as host state, and no redesignation until all party states had served

as host state. In 1983, Illinois opened discussions with Kentucky to form a
two-state compact.

In 1984, legislation was enacted creating the Central Midwest Interstate
LLRW Compact, including Illinois and Kentucky, with Illinois as the host
state (Illinois, State of, P.A. 83-1340, 1984). Similar legislation was passed
in Kentucky. The laws provided that designation of a host state would be for
a maximum of 20 years and provided that no state which generates less than
ten percent of the region's waste could be designated as a host state, although
other member states could volunteer to be a host state. The amount of
LLRW produced per year by Kentucky is shown in Table 20.

Table 20. TOTAL LOW-LEVEL WASTE GENERATED IN KENTUCKY
AND DISPOSED OF AT COMMERCIAL DISPOSAL SITES BY YEAR

	Kentucky	
	cubic meters	curies
1979	194	37
1980	368	5,320
1981	64	107
1982	73	24
1983	74	5
1984	47	4
1985	101	3
1986	107	14
1987	5	40
1988	60	762
1989	285	21
1990	131	61

Source: DOE, 1980 - 1991.

Illinois chose to form a compact because it would make it possible to exclude
out-of-compact waste. As a major generator, Illinois chose to enter into a
compact with Kentucky, which produces less than ten percent of the region's
waste. The agreement was also intended to create a reciprocal arrangement
in case of significant increase in Kentucky's waste production.

The law stated that the purposes of the compact were to: provide the
instrument and framework for a cooperative effort; provide sufficient
facilities for LLRW management in the region; protect the health and safety

of citizens; limit the number of LLRW facilities; promote the volume and source reduction of LLRW; distribute the costs, benefits, and obligations (including liability) of LLRW management equitably among party states and generators; ensure ecological and economical LLRW management, including the prohibition of shallow land burial; and promote the use of above-ground facilities.

The law approving the compact created a three-member Commission to administer the compact. The Commission would include two members from Illinois (because Illinois was the only state that would be designated as a host state), one member from Kentucky, and one non-voting member from each host county. The Compact provided that host states would levy surcharges on users of the regional facility(ies) to cover the Commission's budget.

The Commission may approve the export of waste produced in Illinois or Kentucky to storage, treatment, or disposal facilities outside the region. Both Commission members from the host state must approve the use of a regional facility for waste from outside the compact region. The compact law specifically excluded the waste originating from the Maxey Flats (Kentucky) site from disposal at the regional LLRW facilities.

The law required the Commission to develop a regional management plan that promoted waste reduction and determined the number and type of regional LLRW storage, treatment, and disposal facilities needed. According to the law, the host state will be responsible for planning, siting, operating, monitoring, decommissioning, and providing extended care of the facility. The Central Midwest LLRW Compact was approved by Congress in 1985.

Since its adoption, the Illinois LLRW Management Act has been amended several times. In 1986, language prohibiting the use of shallow land burial as the disposal method was added to the Act (Illinois, State of, P.A. 84-244, 1986). The 1986 amendments also required that multiple, alternative sites be selected and characterized, and authorized grants to local governments where alternative sites are located which could be used to pay for up to 50 percent of the cost of independent review of site suitability.

In 1988, amendments were added that required the approval of the county government if a facility is not located within 1-1/2 miles of any municipality (Illinois, State of, P.A. 85-1133, 1988). These amendments also authorized IDNS to make unrestricted grants which could be used for any lawful purpose to counties or municipalities within whose jurisdiction an alternative site for characterization was located.

After two sites -- the Martinsville (Clark County) and the Geff (Wayne County) sites -- were selected for characterization in 1988, state officers, environmentalists, and affected citizens raised various concerns. IDNS was criticized for moving too fast towards selecting a site without collecting enough data about alternative sites. Later, the Department was accused of making improper changes in the text of a draft characterization report prepared by the siting contractor, Battelle Memorial Institute. These included changes to language approved by the Illinois State Geological and Water Surveys (ISGW Surveys) without their approval. This charge generated considerable media attention and public concern about the safety of the site and the conflicting roles of IDNS as facility developer and regulator.

In late 1989, the Governor asked a former Illinois Supreme Court Justice to hold public hearings and determine whether the proposed sites were safe. Almost simultaneously, the Illinois Senate passed a resolution to investigate the LLRW program. IDNS was ordered to stop all site-selection activities until the investigation was completed (IDNS, 1991). In April 1990, the Director of IDNS resigned.

In May 1990, the staff of the Senate Executive Committee submitted a report of its findings to the Committee. That charged IDNS with attempting to buy local acceptance of sites in Wayne County and Martinsville and understating the potential risk to public health.

Following the Senate investigations, in Spring 1990, the Legislature amended the LLRW Management Act (Illinois, State of, P.A. 86-1044 & P.A. 86-1050, 1990). The amendments decreased the control of IDNS over the site-selection process by establishing an independent LLRW Disposal Facility Siting Commission. The duty of the Commission was to evaluate the safety and suitability of any site proposed by IDNS. Under the new law, IDNS cannot issue a license until after the Siting Commission approves the site.

In June 1990, the Governor appointed a former Illinois Supreme Court Justice, a civil engineering professor, and an environmentalist to the Siting Commission, which was required to hold adjudicative-style public hearings on the suitability of any proposed site in the host county. Hearings on the Martinsville site, which was proposed on January 10, 1991, started in June 1991. In October 1992, the Siting Commission expressed serious concern about the scientific integrity of the project and unanimously rejected the Martinsville site. Over the course of five years, the siting process cost Illinois generators an estimated $85 million.

INSTITUTIONAL STRUCTURE

In response to the Three Mile Island reactor accident in Pennsylvania, the Governor of Illinois created the Illinois Department of Nuclear Safety (IDNS) in April 1980 to protect the public from the potential hazards associated with ionizing radiation (IDNS, 1991).

The LLRW Management Act (Illinois, State of, P.A. 83-991, 1983) made IDNS responsible for regulating and licensing LLRW storage, treatment, and disposal facilities; finding locations for LLRW facilities; collecting fees from

Illinois generators; and providing a mechanism to compensate for any losses caused by releases of radioactivity from a LLRW disposal facility. The law required the ISGW Surveys to cooperate with IDNS in identifying technical considerations relating to the siting of a disposal facility and in performing site characterization and environmental impact studies.

In 1985, IDNS created the Citizens' Advisory Group on Low-Level Radioactive Waste to advise IDNS on public concerns about the disposal facility. The Group included representatives of environmental groups, generators, and other citizens. Later in 1985, IDNS established the Technical Advisory Committee to advise on technical aspects.

In 1985, the Central Midwest Compact Commission was established, with the primary duties of planning waste reduction and management within the compact region and overseeing waste import and export.[44] Members of the Commission included two representatives from Illinois and one representative from Kentucky. By 1989, the most important tasks of the Commission were to identify the LLRW disposal, storage, and treatment needs of the region and to develop a regional LLRW management plan. From 1988 to 1991, the Commission awarded grants to communities where alternative sites for the disposal facility had been identified.

The LLRW Management Act instructed IDNS to select one or more contractors to develop and operate a disposal facility. The developer/operator was required to design, finance, operate, and assume liability for the disposal facility. IDNS would oversee operations at the site and monitor environmental and health impacts.

In November 1987, IDNS sent out requests for proposals for a LLRW disposal facility developer/operator (IDNS, 1987). Two proposals were submitted by Westinghouse Electric Corporation and a group of companies led by Chem-Nuclear Systems, Inc. Proposals were reviewed by IDNS, other experts, and the Citizens' Advisory Group. The proposals were evaluated in terms of the proposer's technical, financial, management, and socioeconomic qualifications; technical merit of the proposal; socioeconomic merit of the proposal, including public involvement plan; financial responsibility and liability; economic merits; and performance assurances (IDNS, 1987).

In 1988, Westinghouse was selected as the developer and entered a contract to design the facility. In May 1989, however, Westinghouse expressed concerns about the requirements on facility financing, operator liability, and facility ownership, and withdrew from the contract. In July 1989, IDNS entered into a contract with Chem-Nuclear Systems, Inc. The company agreed to design, finance, operate, and assume liability for the disposal facility (IDNS, 1991).

---

[44] Central Midwest Interstate LLRW Compact Act (Illinois, State of, P.A. 83-1340, 1984).

In June 1990, the LLRW Management Act was amended to establish a three-member LLRW Disposal Facility Siting Commission to evaluate the suitability of any site recommended by IDNS. The three members of the Siting Commission were appointed by the Governor with the consent of the Senate in June 1990 (IDNS, 1991).

DESIGN OF THE SITE- AND METHOD-SELECTION PROCESSES

The Illinois LLRW Management Act (Illinois, State of, P.A. 83-991, 1983) directed IDNS to: identify geologic and hydrologic conditions and locations best suited for a disposal facility (in cooperation with ISGW Surveys); select at least three alternative sites for characterization; conduct characterization and environmental impact studies (in cooperation with ISGW Surveys); and select a site for the disposal facility.

The Act required public participation, which was mainly in the form of public comment and hearings. In late 1984, IDNS started to develop its own public participation plan.[45]

In 1987, the IDNS hired Battelle Memorial Institute to perform studies for site selection. Environmental groups criticized the IDNS for going ahead with site selection without developing required rules on design and operation of the facility, and threatened to sue the Department. Then IDNS planned a process in which site and method selection would be conducted in parallel, and, in 1988, it hired Westinghouse (replaced in 1989 by Chem-Nuclear) to develop a facility design.

In January 1988, IDNS published a Site Identification Plan and Status Report (IDNS, 1988). According to the plan, the site identification process was to be based on an integrated assessment of the "technical excellence and political acceptability" of areas. IDNS adopted a voluntary process in which counties were allowed to withdraw from consideration at any time in the process. The planned site selection would first process six steps: identify potential candidate areas for reconnaissance studies. Next, eight potential alternative sites (about four square miles) would be selected for preliminary evaluation. Four alternative sites (about four square miles) would then be selected for detailed geotechnical and environmental characterization. The geology and hydrology of the alternative sites would be characterized, and the environmental and socioeconomic impacts of the disposal facility would

---

[45] In addition to those prescribed by law, public participation included advisory groups, including a citizens' advisory group, adhoc panels, and local committees for affected communities; meetings, including public hearings, briefings, workshops, and conferences; informational material, including newsletters and other media releases, studies, and reports; and other mechanisms, including informal review of preliminary drafts of rules.

be assessed. Finally, one site (about 1,000 acres) would be selected for the facility.

Most counties withdrew from consideration early in the site-selection process, so during 1988 IDNS worked to change the LLRW Management Act to require IDNS to select at least two, rather than at least three, alternative sites for characterization. This allowed IDNS to focus on the Martinsville and Geff sites.

In October 1989, the Governor revised the site-selection process to include a completed license application by the facility developer prior to selecting a site.

The 1990 amendments to the LLRW Management Act (Illinois, State of, P.A. 86-1044 & 86-1050, 1990) changed the site-selection process again, specifying that the facility design be considered in determining the suitability of any site. According to the new law, IDNS was to propose a site to the Siting Commission and the Commission was to determine its safety and suitability. IDNS retained responsibility for reviewing the license application submitted by the developer and issuing a proposed license, while the Commission was made responsible for reviewing the proposed license and determining if it were consistent with its earlier review of the facility design.

SITE SELECTION FOR THE PERMANENT DISPOSAL FACILITY

The Illinois LLRW Management Act (Illinois, State of, P.A. 83-991, 1983) prescribed a number of criteria for siting LLRW storage, treatment, or disposal facilities.[46]

In June 1987, Battelle Memorial Institute was selected by IDNS to perform the necessary investigations to identify a site for the disposal facility. As the first step, information was provided to each county about the LLRW management program. Initially 21 counties expressed an interest in learning more about the program.[47]

Next, exclusionary and favorability factors for site selection were defined; favorability criteria were further subdivided into performance-related and non-performance-related groups.[48]

---

[46] These included protection of public health, safety, and welfare; suitable geology and hydrology; minimizing the possibility of radioactive releases into groundwater; minimizing transportation impacts; and excluding areas within 1-1/2 miles of the boundaries of any municipality unless approved by the municipal government.

[47] Bond, Carroll, Cass, Champaign, Clark, Cumberland, Dewitt, Effingham, Fayette, Ford, Gallatin, Greene, Iroquois, Knox, Logan, Marshall, Menard, Peoria, Piatt, Sangamon, and Vermilion counties.

[48] Exclusionary factors included: designated federally-protected lands; designated State-protected lands; and areas with freestanding water; with high earthquake potential; prone to

Exclusionary factors and performance-related favorability factors were applied to the 21 counties that had expressed some interest in the LLRW management program. Areas larger than four square miles having no exclusionary characteristics and no characteristics that were potentially adverse to performance were identified as potential candidate areas. This resulted in the identification of 60 potential candidate areas in 17 of the 21 counties. By the end of January 1988, however, all 21 counties withdrew from participation in the program (IDNS, 1990).

In February 1988, the City Council of Martinsville (Clark County) passed a resolution asking IDNS to consider land near the city. IDNS identified three potential candidate areas located within 1.5 miles of the municipal boundaries of Martinsville. In early 1988, officials of Wayne and Warren counties also requested that their counties be considered. Warren County later withdrew, but Wayne County decided to remain under consideration. IDNS identified 13 potential candidate areas in Wayne County.

In the spring and summer of 1988, the 19 potential candidate areas (six in Clark County and 13 in Wayne County) were analyzed, applying the exclusionary and performance-related favorability criteria followed by reconnaissance IGSW Surveys. On this basis, IDNS selected six candidate areas. These were Martinsville North (Clark County); Martinsville East (Clark County); Martinsville South (Clark County); Geff (Wayne County); Crisp (Wayne County); and Bluford (Wayne County).

In 1988, additional reconnaissance studies were performed in each candidate area to collect more detailed geology and hydrology data. By August 1988, IDNS and its contractors had selected alternative sites in four of the six candidate areas: the Martinsville North and the Martinsville South Alternative Sites (Clark County), and the Geff and Crisp Alternative Sites (Wayne County). At the Crisp site, however, the landowners refused to permit access for characterization studies. To maintain a voluntary process, IDNS decided to drop the Crisp site from further consideration. For technical reasons, the Martinsville South site was also dropped. In October

---

subsidence or landsliding; and within a 100-year floodplain. Favorability factors included areas of low permeability; with simple geologic structure; without surficial sand and gravel deposit; with low erosion; far from surface water supplies; and without high-yield groundwater aquifers (IDNS, 1988).

Non-performance related favorability factors included areas: without known resources with high potential for development and economic recovery; close to federal and State primary roads; without critical habitats of endangered or important species; with as little farmland as possible; that are not critical recharge or discharge zones; without archaeological, historical, or cultural sites; not likely to experience future urban development; without petroleum and chemical plants, fossil and nuclear power plants or existing commercial disposal sites; without utility corridors, rail corridors, or highways bisecting the site; that are close to the generators of LLRW; and areas where a LLRW disposal facility would not create major traffic impacts, particularly on secondary roads.

1988, IDNS designated the Martinsville (North) and the Geff alternative sites for detailed characterization.

In November 1988, advisory referenda were held in both Clark and Wayne Counties on the acceptance of the disposal facility. In both counties, results indicated resistance to hosting the facility, although voters in the City of Martinsville were in favor.

In the summer of 1988, IDNS and the IGSW Surveys began conducting detailed studies of environmental conditions, geological structures, and hydrological systems at the Martinsville and Geff sites.[49] Data collection at the Martinsville site was expanded in the summer of 1989.

In August 1989, IDNS began holding public workshops to discuss draft reports summarizing the results of the characterization studies. In October 1989, the IGSW Surveys publicly withdrew their support for one of the reports because they said that IDNS had made changes to the draft Martinsville Alternative Site Investigation Study Hydrological Investigations report without their concurrence. At a subsequent meeting between IDNS and the IGSW Surveys' staffs, it was agreed that more data would be collected (IDNS, 1991).

In April 1990, after a nonbinding referendum in Wayne County showed little support for the facility, the Wayne County Board announced its intention to deny approval of the Geff site if approval were requested by IDNS. In January 1991, IDNS recommended the Martinsville site to the Siting Commission as the proposed site for the LLRW disposal facility.

After the amendments to the LLRW Management Act were passed in June 1990, an independent LLRW Disposal Facility Siting Commission had been established and held its first meeting in August 1990. The Commission started hearings on the Martinsville site in the summer of 1991. Participants in the hearings included IDNS and its contractors, elected officials, citizens, technical and legal experts, and various groups and organizations of Clark County and the City of Martinsville. The County government opposed the siting, while the City favored it. One local group, the Concerned Citizens of Clark County, opposed the facility, whereas another, the People for Responsible Opportunities, was supportive.

Public hearings were concluded in March 1992. The Siting Commission, however, postponed making its determination of the safety of the proposed facility until the Supreme Court decided New York State's challenge to the LLRW Policy Amendments Act (*The Radioactive Exchange*, Vol. 11, Nos. 8 & 9, 1992).

In October 1992, the Siting Commission voted 3-0 against the Martinsville site. The Commission claimed that IDNS officials made

---

[49] Data were collected about the ecology, soils, land use, meteorology and air quality, water resources, cultural resources, noise, socioeconomics, transportation, geological and geotechnical characteristics, and surface and groundwater conditions at both sites.

fundamental errors in gathering data to support the case.   Unanswered scientific questions included whether sand beneath the proposed facility was linked to the town's water supply, and whether the facility would survive for 500 years.

PERMANENT DISPOSAL METHOD SELECTION

The LLRW Management Act of 1983 prescribed choosing "the best available management technologies which are economically reasonable, technologically feasible and environmentally sound" (Illinois, State of, P.A. 83-991, 1983). The Central Midwest Interstate LLRW Compact Act (Illinois, State of, P.A. 83-1340, 1984) also had excluded shallow land burial from consideration and proposed "the use of above-ground facilities and other disposal technologies providing greater and safer confinement of low-level radioactive waste than shallow-land burial facilities."   The law instructed the Compact Commission to develop alternative waste management technologies.   In order to be in accord with the compact law, the LLRW Management Act Amendments of 1986 (Illinois, State of, P.A. 84-244, 1986) also prohibited the use of shallow land burial as the disposal technology.

In 1986 IDNS promulgated regulations on the design, construction, and operation of the disposal facility (Illinois, State of, Adm. Code Title 32, Part 606, 1986).   These regulations mandate a one millirem per year exposure limit for the public from external gamma radiation during operations, a substantially more stringent standard than the 25 millirem per year exposure limit established by the NRC Part 61 regulations.

The Regional Management Plan prepared by the Compact and its contractors (Central Midwest Interstate LLRW Compact, 1987) investigated several disposal technologies: below-ground vaults; above-ground vaults; modular concrete canisters; earth-mounded concrete bunkers; augured holes; and mined cavities.

The study evaluated the various disposal technologies in terms of protecting public health and safety; protecting intruders; protecting workers; cost; and licensability (Central Midwest Interstate LLRW Compact, 1987).

The six technologies were rank-ordered using the five criteria (Table 21). No overall evaluation was performed, and the Commission did not make specific recommendations about the preferred disposal technology.   They concluded that "the Central Midwest regional disposal facility will probably incorporate some of the best features of several of the designs that were discussed" (Central Midwest Interstate LLRW Compact, 1987).

Table 21. RANK ORDERING OF DISPOSAL TECHNOLOGIES

Ranking Criteria

	Protecting Public Health and Safety	Protecting Intruders	Protecting Workers	Cost	Licensability
MOST FAVORABLE	MC	MC	AH	[AH BGV MCCD]	[BGV MCCD EMCB AH]
↕	[BGV MCCD EMCB AH]	[BGV MCCD EMCB AH]	[BGV MCCD EMCB AH]	AGV EMCB MC	AGV MC
LEAST FAVORABLE	AGV	AGV	MC		

LEGEND:

BVG = Belowground vaults
AVG = Aboveground vaults
MCCD = Modular concrete canister disposal
EMCB = Earth mounded concrete bunkers
AH = Augered holes
MC = Mined cavities

Source: Central Midwest Interstate LLRW Compact, 1987

In its Request for Proposals to Design, Operate, and Close a Low-Level Radioactive Waste Facility (IDNS, 1987), IDNS did not prescribe particular disposal technology, but specified that the facility design should: minimize the possibility of release; maximize monitoring efficiency and sensitivity; enhance remedial actions, including the removal of waste; and complement the natural protection afforded by careful site selection.

In 1988, Westinghouse was selected as site developer and was required to develop a basic design for the facility. Westinghouse proposed above-grade vaults with modular concrete canisters.

In 1989, after Westinghouse had withdrawn from the contract, Chem-Nuclear developed a preliminary facility design (Chem-Nuclear Systems, Inc., 1989). This design included above-ground vaults containing modular concrete canisters, with multi-layered vault covers and monitoring devices. In May 1991, Chem-Nuclear submitted the license application to IDNS (Chem-Nuclear Systems, Inc., 1991/a). In early 1992, the license application was reviewed by the licensing staff of the Department, and in April 1992, Chem-Nuclear submitted its responses to questions posed by IDNS on the license application.

Illinois State law includes several provisions concerning host community monitoring and oversight of disposal facility operations. The LLRW Management Act Amendments of 1988 (Illinois, State of, P.A. 85-1133, 1988) give the host community governing body the statutory power to close the facility if it accepts any waste except LLRW or mixed LLRW for disposal. IDNS was negotiating with the host community on additional safeguards and independent monitoring activities at the disposal facility.

In June 1992, the City of Martinsville unanimously approved a "resolution of support" and a community agreement with IDNS. Under the community agreement, the City would have full access to the facility and its records, would be allowed at least one inspector at the site at all times, and would have the right to shut down the facility if certain regulations were violated.

COMPENSATION AND INCENTIVES

The LLRW Management Act of 1983 (Illinois, State of, P.A. 83-991, 1983) directed IDNS to adopt regulations pertaining to compensation for losses due to release of radioactivity from the disposal facility. The law also provided that generator fees could be appropriated for payment in lieu of taxes to the host community.

In 1986, amendments were added to the LLRW Management Act (Illinois, State of, P.A. 84-244, 1986) providing for matching grants to local governments where alternative sites for the disposal facility are located.

These grants enabled affected communities to independently review site characterizations and related studies. The grants were to be funded by generator fees and could be as great as 50 percent of the local governments' costs for such studies, with a maximum grant of $50,000.

After the Martinsville City Council and the Wayne County Board voted in favor of participating in the siting process in Spring 1988, both governments were granted $100,000 from the Compact Commission and $50,000 from IDNS to review the siting studies and mitigating impacts on their community.[50]

The amendments of 1988 (Illinois, State of, P.A. 85-1133, 1988) deleted matching requirements and grant limits from the Management Act. In addition, the amendments authorized IDNS to award unrestricted grants to counties and municipalities where an alternative site is located. Between 1989 and 1992, the City of Martinsville received $400,000 each year in unrestricted grants from IDNS and $250,000 each year in restricted grants from the Compact Commission. While under consideration as a potential host community, Wayne County also received a total of $950,000 from IDNS and the Compact Commission.

The community ultimately selected to host the site is to receive annual compensation from waste disposal surcharges during the operating life of the facility. The amount of this compensation was determined through negotiations between IDNS, Chem-Nuclear, and the community.

According to the agreement between the community and Chem-Nuclear, the firm would pay the city $2.15 million a year for the duration of operation. Chem-Nuclear would also pay a fee in-lieu-of-taxes. That fee would start at about $500,000 in the first year and could reach $2.2 million in ten years. In addition, the firm guaranteed 200 construction jobs and 100 permanent jobs. The facility was designed to operate for 50 years. (*The Radioactive Exchange*, Vol. 11, No. 10, 1992). The Martinsville site, however, was rejected by the Siting Commission in October 1992.

INTERIM WASTE MANAGEMENT

The LLRW Management Act of 1983 (Illinois, State of, P.A. 83-991, 1983) required IDNS to develop plans for interim management of Illinois' waste after January 1, 1986, until a permanent disposal facility is operational.[51]

---

[50] The grants were restricted to these purposes.

[51] IDNS investigated four options for interim storage: Non-reactor generators continue to ship their waste to the existing out-of-state facilities, while reactor generators would ship as much as possible and store the remainder on-site; a central interim storage facility is established for non-utility generators; utilities store non-utility waste on an emergency basis; and all generators store on-site (IDNS, 1985/a).

The IDNS study (IDNS, 1985/a) recommended that all generators continue to ship as much waste as possible to out-of-State facilities, while reactor generators store some of their waste on-site, if feasible.

For some years, the issue of interim storage did not attract much attention because it was assumed that Illinois would have a disposal facility by January 1993 in accordance with the deadlines of the federal LLRW Policy Amendments Act of 1985 . Due to delays in the siting process, however, in the spring of 1991, IDNS conducted an interim storage survey with the generators of Illinois and Kentucky. The survey found that generators and brokers have sufficient storage capacity for the calendar years 1993 and 1994.

ANALYZING THE ILLINOIS CASE HISTORY FROM THE PERSPECTIVE OF THE COMPETING VALUES APPROACH

The Illinois siting process was initially highly goal-directed, but the process ultimately proved to be inefficient; it was comparatively quite expensive without producing a solution to the State's LLRW management problem.

The 1983 LLRW Management Act made IDNS responsible for regulating and licensing LLRW facilities, site and method selection, and compensation. After forming an interstate compact with Kentucky in 1985, Illinois made serious efforts to site a disposal facility for the LLRW generated in the two states. The site-selection process progressed briskly until 1988, when public concerns about the safety of the selected site and IDNS's conflicting joint roles as both developer and generator surfaced, eventually leading the Legislature to mandate significant changes in the institutional structure. An independent Siting Commission was established in 1990 to evaluate the suitability of the selected Martinsville site. In 1992, the Commission found the site unsuitable, thus blocking the licensing procedure.

The siting process was relatively expensive partially because substantial incentives were provided for communities and counties willing to be considered potential hosts of the facility. Between 1987 and 1992, the siting process cost Illinois generators approximately $85 million.

Many of the difficulties associated with the Illinois process appear to have arisen from its perceived shortcomings from the empirical perspective, which emphasizes data-based processes and accountability in decisions. In the early stages of the process, IDNS seemed to place relatively little emphasis on data-collection and accountability. Subsequently, the agency was criticized repeatedly on data-related counts. In 1987, environmental groups threatened to sue IDNS for starting site selection without developing rules for facility design and operation. In 1988, state officers, environmentalists, and affected citizens criticized IDNS for moving too fast with site selection without collecting enough data about alternative sites. In 1989, IDNS was accused of making unauthorized and unwarranted changes in the text of a draft characterization report prepared by its siting contractor.

The 1991 amendments to the LLRW Management Act led to a series of public hearings investigating the quality of the data collected at the Martinsville site, which concluded that IDNS made fundamental errors in data collection. The inability to provide sufficient accountability for the selection of the Martinsville site appears to have been a key element contributing to its eventual rejection.

As for the consensual perspective, which emphasizes participatory processes and supportable decisions, the types of public participation mandated by Illinois law consisted mainly of public comments and hearings. In 1984, IDNS developed a public participation plan that included a Citizens' Advisory Group, local committees for affected communities, and extensive public information and education programs. In 1985, a Citizens' Advisory Group including representatives of environmental groups, generators, and other citizens, was formed to advise IDNS. The Group had only review power and little impact on the siting process, however.

In 1988, IDNS adapted a voluntary siting process in which counties and municipalities were allowed to withdraw from being considered as hosts at any time. The host community would also have the right to monitor the facility and close it down if certain regulations were violated. Compensation and incentives awarded to the host community were defined through negotiations. The Illinois process granted comparatively high levels of decision power to local governments.

Regarding the political perspective, which emphasizes adapability and political legitimacy, the volunteer process adapted in Illinois was comparatively quite flexible, especially concerning interactions with local communities. IDNS, Chem-Nuclear, and local governments conducted direct negotiations about a number of key issues, including facility oversight, compensation, and incentives. IDNS's perceived lack of responsiveness to concerns about the data quality and site safety and suitability seems to have contributed, however, to the lack of political legitimacy accorded to the site selection decision.

Chapter 6

NEBRASKA

BACKGROUND

In 1979, approximately 801 cubic meters (104 curies) of LLRW was generated in Nebraska and disposed of out-of-state. Virtually all of this waste was produced by the two publicly-owned nuclear power plants in the State. As shown in Table 22, the amount and radioactivity of LLRW generated each year in the state has fluctuated. By 1990, however, the volume of LLRW produced per year had been considerably reduced from earlier levels (DOE, 1980, 1981, 1982, 1983, 1984/a, 1985, 1986/a, 1987, 1988, 1989, 1990, 1991).

Table 22. TOTAL LOW-LEVEL WASTE GENERATED IN NEBRASKA AND DISPOSED OF AT COMMERCIAL DISPOSAL SITES BY YEAR

	Volume (cubic meters)	Radioactivity (curies)	Volume Percentage by Source		
			Institutional and Industrial	Power Reactor	Government and Military
1979	801	104	<1%	99%	0%
1980	836	1,869	1%	99%	0%
1981	751	600	1%	99%	0%
1982	827	593	5%	95%	0%
1983	992	1,402	0%	100%	0%
1984	871	727	6%	94%	0%
1985	1,060	30,012	1%	99%	0%
1986	576	625	1%	99%	0%
1987	493	3,090	0%	100%	0%
1988	363	174	1%	99%	0%
1989	473	328	1%	99%	0%
1990	441	378	2%	98%	0%

Source: DOE, 1980 - 1991.

A chronological summary of key events related to the LLRW disposal facility siting in Nebraska is given in Table 23.

Table 23. CHRONOLOGICAL SUMMARY OF KEY EVENTS IN NEBRASKA

Date	Event
1980	Nebraska created a Citizens' Advisory Board on LLRW.
1981	Nebraska began negotiations with representatives of two potential compact regions (Central Interstate and Midwest).
1982	The Nebraska Legislature passed Resolution 233 supporting Nebraska's participation in the Central Interstate Compact.
1983	Arkansas, Kansas, Louisiana, Nebraska, and Oklahoma formed the Central Interstate LLRW Compact.
1984	The Compact Commission hired Dames and Moore to conduct a site-exclusionary study of the region.
1985	Congress ratified the Central Interstate LLRW Compact. The Phase 1 Exclusionary Study, which identified candidate areas in each of the states, was completed and released.
1986	The Nebraska Legislature passed Legislative Bill (LB) 491 adopting the LLRW Disposal Act.
1987	The Phase II Exclusionary Study was completed. LB 426, prohibiting shallow land burial, was adopted. The Compact selected US Ecology to develop, construct, and operate a LLRW disposal facility. The Compact Commission selected Nebraska as the host state.
1988	The 402 Initiative, calling for withdrawal from the Compact, was defeated in a Statewide referendum. LB 1092, specifying the State's policy regarding the LLRW issue, was adopted. Twenty counties formally expressed interest in the screening process to US Ecology. The village of Butte in Boyd County submitted a letter of interest to US Ecology. US Ecology began screening the siting areas to locate available land.
1989	Three candidate sites were selected by US Ecology. A site near Butte in Boyd County was selected by US Ecology.
1990	US Ecology submitted an application for a license to construct and operate a LLRW disposal facility at the Butte site. Boyd County residents protested against the site.
1991	LB 838, requiring shared liability by the other Compact states, was passed by the Nebraska Legislature. The Director of the Compact Commission was arrested and later convicted for embezzling Compact funds. A GAO report concluded that the characterization of the three candidate sites appeared to have been done in a technically correct manner. DEC and DOH certified that the license application was complete. Voters in McCulley Township, where the site was located, passed regulations prohibiting the storage of radioactive materials within the Town.

OVERALL APPROACH

The State of Nebraska responded to the 1980 federal Low-Level Radioactive Waste Policy Act by creating a Citizens' Advisory Board on LLRW. In 1981, Nebraskan representatives began negotiations with representatives of two potential compact regions (Central Interstate and Midwest). They decided to join the Central Interstate Compact, partly because the states associated with this compact generated less LLRW than states associated with the Midwest Compact. Joining the Central Interstate Compact, instead of the Midwest Compact, placed a greater financial burden on Nebraska for project development. In March 1982, the Nebraska Legislature passed Resolution 233 supporting Nebraska's participation in the Central Interstate Compact. Subsequently, during 1982, ten public meetings across the State were held by State legislators and the Nebraska Department of Environmental Control (DEC) to assess what Nebraskans felt the state should do to ensure future disposal capacity for the State's LLRW (US Ecology, 1988; 1990).

By the spring of 1983, five states, Arkansas, Kansas, Louisiana, Nebraska, and Oklahoma, had ratified the Central Interstate Low-Level Radioactive Waste Compact. The amount of waste produced per year by the other states in the Compact is shown in Table 24. The purpose of the Compact was to provide a framework for cooperative effort among the states in the region; to promote the health, safety, and welfare of both the citizens and environment of the region; to limit the number of facilities needed to manage LLRW; and to distribute the costs, benefits, and obligations among the party states. According to the Compact, all member states are required to serve as a host state in turn (a facility is expected to operate for 30 years). In addition, a Compact Commission was established to oversee the responsibilities of the Compact (Nebraska, State of, LB 200, 1983).

Nebraska joined the Compact by passing LB 200 (Nebraska, State of, LB 200, 1983). In 1985, Congress ratified the Central Interstate Low-Level Radioactive Waste Compact (Central Interstate LLRW Compact Commission, 1990; US Ecology, 1988; 1990).

In 1984, the Compact Commission hired Dames and Moore, an environmental consulting firm, to conduct a site-exclusionary study of the region. The study was designed to identify general areas of each member state that were unsuitable for a shallow land burial LLRW disposal facility. The Phase 1 Exclusionary Study, which identified candidate areas in each of the states, was completed and released in June 1985. A second phase of the exclusionary study was conducted to locate potential siting areas within each of the candidate areas. Potential siting areas were identified with site characteristics suitable to meet the U.S. Nuclear Regulatory Commission's (NRC) siting criteria for Licensing Requirements for Land Disposal of

Table 24. TOTAL LOW-LEVEL WASTE GENERATED IN THE NON-HOST STATES OF THE CENTRAL COMPACT AND DISPOSED OF AT COMMERCIAL DISPOSAL SITES BY YEAR

	Arkansas		Kansas		Louisiana		Oklahoma	
	cubic meters	curies	cubic meters	curies	cubic meters	curies	cubic meters	curies
1979	265	180	10	1	19	1	21	266
1980	251	1,256	17	<1	<1	3	71	<1
1981	1,566	4,196	47	1	2	<1	176	<1
1982	839	3,314	14	0	30	10	103	20
1983	790	2,056	0	0	16	3	64	700
1984	947	1,375	39	0	14	0	140	1
1985	720	1,864	48	170	311	34	298	352
1986	127	221	186	403	638	117	1,411	3
1987	556	1,529	128	119	788	522	2,346	13
1988	199	823	142	1,472	521	4,787	806	233
1989	226	322	180	2,471	615	1,478	908	4
1990	177	14	321	67	92	994	622	3

Source: DOE, 1980 - 1991.

Radioactive Waste. The Phase II Exclusionary Study was completed in August 1987 (Central Interstate LLRW Compact Commission, 1990).

In 1986, the Nebraska Legislature passed LB 491 (Nebraska, State of, LB 491, 1986) which adopted the Low-Level Radioactive Waste Disposal Act. Through this law, the Legislature indicated that it was the policy of the State of Nebraska, in furtherance of its responsibilities to the Central Interstate Low-Level Radioactive Waste Compact Commission and to protect the state's citizens and environment, to provide for the capacity within or outside the State for the commercial disposal of LLRW. The law further specified that it was the policy of the State that the cost of the disposal of LLRW be borne by the generators. The Act specified that the Department of Environmental Control should be the lead agency in Nebraska and that volunteer host communities should be sought (Nebraska, State of, LB 491, 1986).

Before a host state was designated, another legislative bill (LB 426) was introduced to modify the LLRW Disposal Act of the previous year. In this bill, the use of shallow land burial as the disposal method was forbidden. In May 1987, this bill was adopted into law (Nebraska, State of, LB 426, 1987). At the same time, LB 427, mandating Nebraska's withdrawal from the compact was introduced, but was not adopted.

The Compact Commission then solicited proposals to hire a contractor to develop, construct, and operate a LLRW disposal facility. Proposals were submitted by US Ecology and the Westinghouse Corporation in April 1987. Typically, such proposals are submitted as confidential documents. The Compact Commission, however, requested that each bidder file sufficient copies to be placed in public reading rooms in each member state. The competing bidders were also asked to appear before public technical advisory panels in each state for public questioning. Finally, the bidders were required to undertake a three-week tour of the five-state region, appearing at a series of public meetings arranged by the Compact Commission. Attendance at these meetings varied from a handful of people to several thousand. Two such meetings were held in Nebraska. On June 29, 1987, the Compact selected US Ecology based on their corporate experience and history, financial commitment, design proposal, and program for public involvement (Central Interstate LLRW Compact Commission, 1990; GAO, 1991; US Ecology, 1988; 1991). US Ecology began work on the project soon after selection, but contracts between the contractor and the Compact Commission were not finalized until January 1988.

Dames and Moore was also hired to develop the Regional Management Plan. As part of this plan, Dames and Moore assessed the characteristics of waste sources in the region, evaluated alternative disposal technologies, developed a public involvement plan, provided guidelines for submitting a proposal to develop a regional LLRW disposal facility, and established procedures for states that might want to volunteer to host a site. This plan was adopted by the Compact Commission in August 1987 (Central Interstate LLRW Compact Commission, 1990).

As its first task, US Ecology assessed the member states to determine the most suitable host state. Three criteria were used to select the host state: environmental suitability; waste generated; and transportation (US Ecology, 1988).

In evaluating environmental suitability, the Commission instructed US Ecology to incorporate information from Dames and Moore's Phase 1 Report. In evaluating waste generated, US Ecology considered only the number and design type of nuclear power plants in each state. Nebraska ranked highest among member states in both environmental suitability and waste generated. Nebraska's northernmost location in the Compact region ranked the state low in terms of transportation. In December 1987, the Compact Commission selected Nebraska as the host state by a 4-1 vote. Nebraska voted against being selected (US Ecology, 1988; 1991).

After Nebraska was selected as the host state, Governor Kay Orr entered into negotiations with California to allow Nebraska use of the facility proposed for construction in that state. These negotiations were unsuccessful (US Ecology, 1991). At the same time, the Nebraska Right to Vote Committee announced a petition drive that would require community consent and called for withdrawal from the Compact. The 402 Initiative, which appeared on the State ballot in the November 1988 election, was defeated by a vote of 414,394 (65 percent) to 225,174 (35 percent). As a result, Nebraska remained in the Compact.

When Nebraska was selected as the host state, Governor Orr announced ten conditions that must be met if a LLRW disposal facility is to be constructed in Nebraska.[52] In January 1988, LB 1092 was introduced in the Nebraska Legislature. This bill, specifying the State's policy regarding the LLRW issue, incorporated Governor Orr's ten conditions. The bill

---

[52] Governor Orr required that assurance must be provided that US Ecology and the Compact Commission will not locate a facility in a community without community consent; US Ecology must agree to defray the reasonable costs incurred by a local monitoring committee; Nebraska must have complete control over the facility design, the location of the Compact offices, and a veto over the import or export of LLRW into or out of the region; Nebraska must have the right to refuse decommissioning waste; Class C waste must be stored in an easily retrievable form for 30 years, preserving an option to transfer it to a high-level repository if the material is reclassified; mixed waste must be treated to the maximum extent before shipment to the facility; compensation at acceptable levels must be guaranteed to the State and to the host community, paid annually during the operating life of the facility; the State must be completely reimbursed for all costs incurred in regulating, licensing, and planning for the facility; compensation at acceptable levels must be guaranteed to local communities who become active participants in the site-selection process (preoperational compensation would be the obligation of the four other states); and property values in the area surrounding the site must be guaranteed (Governor Orr, undated).

specifically stated that "it is the intent of the Legislature that potential host communities be actively and voluntarily involved in the siting process" and that the "developer shall make every effort to locate the facility where community support is evident" (Nebraska, State of, LB 1092, 1988).

This legislative bill also established requirements regarding site selection and the development of Local Monitoring Committees (LMC). The bill mandated that the developer send written notification to the Governor and Legislature of the selection of three proposed sites by January 1, 1989. Within 30 days after such notification, a LMC had to be established for each proposed site area. The three LMCs would exist only until a site was selected. The LMC operating at the selected site would continue to exist. The role of the LMCs was to "represent the citizens of the proposed site areas and maintain communication with the developer and the department to assure protection of public health and safety and the protection of the air, land, and water resources of the area" (Nebraska, State of, LB 1092, 1988).

An attempt was made to amend LB 1092 to require a vote of the county as evidence of meeting the community consent requirement. This amendment was defeated, but the bill was passed in May 1988.

Early in 1989, LB 761, which was drafted to amend the Nebraska Low-Level Radioactive Waste Disposal Act, was introduced. This legislative bill required that the Compact Commission member from Nebraska be appointed by the Governor with the approval of the majority of the Legislature. The bill called for regular testing of water and agricultural products of landowners adjacent to the site by the Department of Environmental Control and for a legislative study, to be performed by the Judiciary Committee, on liability issues related to the disposal of LLRW (Nebraska, State of, LB 761, 1989). Two attempts were made to amend LB 761 to include a vote by the county as evidence of meeting the community consent requirement. Neither attempt succeeded. In May 1989, LB 761 passed.

In 1990, Ben Nelson ran against Governor Orr in the general election. As part of his campaign platform, Nelson vowed that he would do everything possible to keep the LLRW disposal facility from being built in Nebraska. Opponents to the facility supported his campaign, erecting signs and staging demonstrations against Governor Orr. A rancher in Boyd County, where the site for the facility had been selected, staged a 30-day hunger strike to protest Governor Orr's support of the facility. Upon ending his fast, the rancher endorsed Ben Nelson and made several campaign appearances for him. In November 1990, Ben Nelson was elected Governor of Nebraska.

In January 1991, Governor Nelson introduced two legislative bills, LB 837 and LB 838, which require shared liability among the compact member states (the Compact Commission argues that shared liability had already been agreed to by member states), an additional voting and one non-voting Compact Commission member for the host state, and open meeting and open record provisions for the Compact Commission. LB 837 required that all costs or liabilities incurred by the facility be shared proportionately (according to volume of waste disposed) among the Compact states. According to the bill, if a state does not pass the shared liability provision, they can be denied access to the disposal facility (Nebraska, State of, LB 837, 1991/a). LB 837 was amended to incorporate the provisions of LB 838 and, in March 1991, the combined bill, LB 838, was passed by the Legislature.

The Nebraska law on shared liability requests that the other Compact states adopt Nebraska's shared liability bill as an amendment to the Compact agreement. As of the end of 1992, Arkansas, Louisiana, and Oklahoma had passed such bills, while Kansas had not.

Also in January 1991, attempts were made to amend the community support provisions of LB 1092. LB 72 was introduced into the Legislature. As initially proposed, LB 72 would have required, as evidence of community consent, approval by a majority of the registered voters of the Natural Resource District where the facility is to be located. The bill was later amended to require a majority vote of county residents.

Supporters of the siting process argued that community consent as outlined in LB 1092 (Nebraska, State of, LB 1092, 1988) had been achieved during the siting process and presented as evidence the resolutions of support from the community closest to the site (Butte) and the results from Butte precinct for a number of local and state elections, in which candidates opposed to the site had lost. LB 72 was not adopted.

In April 1991, the Director of the Compact Commission was arrested for embezzling Compact funds (*Omaha World Herald*, June 25, 1991). In response, the Omaha Public Power District and the Nebraska Public Power District, the two utility generators in the State, refused to continue financing the compact unless the Commission allowed generators to audit the Commission; obtain fidelity bonds for the executive director of the Commission; and investigate and recover missing or misappropriated funds. As a result, the Compact Commission instituted several changes in its contract with the six major nuclear waste generators to assure fiscal accountability and oversight (GAO, 1991).

In January 1992, having already been convicted, the ex-Director was sentenced to 50 months in prison. After his sentencing, he made several allegations regarding the way decisions on the LLRW disposal facility were made. The ex-Director alleged that Nebraska volunteered to be the host state and that the Nebraskan Compact Commissioner, the Project Director for US Ecology, and he had worked together to rewrite the Compact selection criteria to get Nebraska selected (*Omaha World Herald*, January 25, 1992). Both former government officials and US Ecology representatives have denied these charges.

The ex-Director also alleged that the Boyd County site had been chosen for political reasons at a closed meeting of Compact officials and US Ecology representatives in October 1989 (two months before the site selection was made public). According to the ex-Director, at that meeting US Ecology officials announced that one other site under consideration was unsuitable because its geology was too complex and that the Commissioner stated that another site was politically unacceptable to then-Governor Orr. This left the Boyd County site as the only remaining alternative.

If this allegation is true, US Ecology could be liable for damages, particularly if the Boyd County site is unlicensable. According to the Nebraska Attorney General, however, there seems to be no clear evidence that the Boyd County site was chosen for political reasons. The Attorney General also noted that the allegations the ex-Director made were statements to reporters that directly contradicted statements he had previously made under oath. Furthermore, the U.S. General Accounting Office conducted an inquiry into the site-selection process at the request of Nebraska U.S. Senator Exon. In that report, dated July 5, 1991, the GAO concluded that the characterization of the three candidate sites appeared to have been done in a technically correct manner and that independent geologists hired by the three communities under assessment all agreed that selection of the Boyd County site was proper (GAO, 1991).

In the spring of 1992, the Compact Commission drafted a compact amendment to assess major generators $16.2 million to continue financing the licensing process for the Boyd County facility. The Boyd County Monitoring Committee filed for an injunction against the Compact Commission because the Committee did not receive sufficient information on the proposed amendment. The U.S. District Court granted a temporary injunction ordering the Commission to provide more background information. The Committee filed for modification, contending the Commission still has not provided sufficient information regarding the underlying basis for the proposed amendment, but this motion was denied. The Compact Commission approved the compact amendment in April 1992. (*The Radioactive Exchange*, Vol. 11, Nos. 8 & 9, 1992).

In August 1992, public opposition to the proposed site escalated to a new level when civil disobedience activities started in Boyd County. The activities were similar to those in New York State in 1990, and New York State environmentalists assisted Nebraska citizen groups (*The Radioactive Exchange*, Vol. 11, No. 16, 1992).

In January 1993, the Nebraska DEC and Department of Health announced their intent to deny the license application because the proposed site included wetlands. In October 1993, these regulatory agencies reversed their position, after U.S. Ecology reconfigured the site to omit the wetlands. Meanwhile, a suit to block site use, which had been filed in federal court by the Governor, was also dismissed.

INSTITUTIONAL STRUCTURE

The Central Interstate Low-Level Radioactive Waste Compact Commission was formed by ratification of the five member states in 1983. Each state has one voting member appointed by the appropriate authority in the state and is responsible for the expenses of its member of the Commission. The Commission was authorized to employ a staff sufficient to carry out its responsibilities and to contract with any person to perform necessary functions to assist the Commission. Initial funding was provided by the states until surcharges from generators could be collected (Central Interstate LLRW Compact Commission, 1990; Nebraska, State of, LB 200, 1983; US Ecology, 1988; 1990).

The Compact Commission was given responsibility for developing policies regarding waste disposal within the compact; providing liaison functions among generators and compact states; selecting a developer (contractor); administering the developer's contract; selecting a host state; managing the developer's actions; disseminating information on project activities; and providing pre-operational funding to regulator and local communities (Nebraska, State of, DEC, undated).

In 1987, the Compact Commission selected US Ecology to serve as the LLRW disposal facility developer (Central Interstate LLRW Compact Commission, 1990). US Ecology was given responsibility for conducting site characterization studies on proposed sites; soliciting public participation; providing pre-operational funding to Local Monitoring Committees; selecting a final site; preparing technical studies for the selected site; preparing the license application; constructing the facility; operating the facility; performing environmental monitoring; and closing the facility (Nebraska, State of, DEC, undated).

The DEC was designated by the Nebraska Legislature as the lead agency regarding the disposal of LLRW in Nebraska. DEC was given responsibility for administering and enforcing statutory requirements; administering funds appropriately; monitoring site characterization work; accepting a license application for review; preparing a safety evaluation report and an environmental impact analysis; conducting public hearings on the license application; developing the terms and conditions of license; making a decision on issuance of license; enforcing the terms of license; performing environmental monitoring; and maintaining the site after closure (Nebraska, State of, DEC, undated).

The Nebraska Department of Health also has responsibilities relating to radiological health issues and had to approve the facility license.

In April 1988, US Ecology gave the Nebraska League of Women Voters a $50,000 grant to establish a Statewide Citizens' Advisory Committee. The League of Women Voters agreed to establish the Committee on the condition that it would be an independent body that would ensure a thorough examination of the issues relating to its designated role of recommending factors for a safe, environmentally sound site for the LLRW disposal facility. To assure credibility, Committee members were selected to represent a wide variety of interests, backgrounds and geographical areas (League of Women Voters of Nebraska, 1989).

Neither the League of Women Voters nor US Ecology had a role in naming specific members to the CAC. Instead, a list of Statewide interest groups and organizations was agreed on by the League and US Ecology.[53] Eight organizations were invited to nominate a citizen(s) to serve on the CAC. Any individual could be nominated; however, it was clearly understood that views expressed at meetings would not necessarily reflect organization positions. The Nebraska County Officials Association, the League of Municipalities, Natural Resource Districts. and Farm Organizations nominated two members (one from the eastern part of the State and the other from the west) for the CAC. The Nebraska Sierra Club, the Nebraska Medical Association, the Nebraska Academy of Science, and the Nebraska Association of Commerce and Industry each nominated one member for the Committee (League of Women Voters of Nebraska, 1989).

There was some controversy regarding membership on the CAC. For example, when asked to assist in the formation of the CAC, the Nebraska Audubon Society refused to nominate a representative. The Nebraska Sierra Club agreed to participate, first recommending a woman who was actively involved with the Statewide group opposing the siting process, Nebraskans for the Right to Vote. She declined, stating that she felt serving on the CAC would be a conflict. The Sierra Club nominated an alternative representative, who agreed to serve. Nebraskans for the Right to Vote criticized the process because they were not allowed to name someone directly to the CAC. They also objected to a retired assistant general manager for a Nebraska utility being nominated to the CAC by the Association of Commerce and Industry (*Omaha World Herald*, March 23, 1988; April 8, 1988; April 9, 1988; April 11, 1988).

When the three candidate sites were announced in January 1989, the Citizens' Advisory Committee completed its work. In accordance with State law, three Local Monitoring Committees (LMC), one in each county where a potential site was located, were established. LB 1092 states that "it is the

---

[53] Single-issue groups and organizations less than one year old were excluded.

intent of the Legislature that the local monitoring committees provide significant input concerning the local needs and resources regarding all relevant aspects of the site selection and, after a site is selected, that the remaining local monitoring committee provide significant input ... regarding all relevant aspects of the construction, operation, monitoring, closure, and custodial care of the facility."

The LMC's were given no regulatory authority. Instead, they were provided with authority and funding to serve as a clearinghouse of information for the LLRW project; to determine local interests and needs as they pertain to waste site planning, construction, and maintenance; to forward information regarding local interests and needs to the developer and regulators; to conduct social and economic impact studies, if desired; to hire consultants to serve as technical staff to the Committee, if desired; to review the license application and provide input to the regulator regarding local needs; and to hire a site inspector during construction and operation phases, if desired (Nebraska, State of, LB 1092, 1988; US Ecology, 1991). According to Nebraska law, the LMCs are to be composed of ten members, all of whom shall be residents of Nebraska.[54]

DESIGN OF THE SITE- AND METHOD-SELECTION PROCESSES

In designing the site-selection process, US Ecology identified five major activities: defining site selection considerations and establishing criteria; screening to identify potential areas; evaluating potential areas to identify potential sites; comparing potential sites; and selecting candidate sites. Screening would be conducted in four stages: screening the State at a 1:500,000-scale level; screening the remaining regions at 1:250,000 scale level; obtaining formal expressions of interest from host areas wishing to participate; and screening interested host entities at a 1:24,000-scale level (US Ecology, 1988).

---

[54] Two members are to be appointed by municipalities which have zoning jurisdiction within 15 miles of the proposed site, or if no municipalities exist, from the municipality in closest proximity to the site to be appointed by the Chief Executive Officer of each municipality or by the Governing Body if there is not a Chief Executive Officer. Two members are to be appointed by the county board. One county board appointee must own real property within a three-mile radius of the proposed site. Two members are to be appointed by the Natural Resources District in which the facility is located. Four members are to be gubernatorial appointees. The gubernatorial appointees are to represent conservation, agriculture, chief of a fire department located within 15 miles of the proposed facility, and an at-large member. All gubernatorial appointees must reside within 15 miles of the proposed facility site (Nebraska, State of, 1990/a).

The general design of the facility (an above-ground, vault-type structure) was proposed before Nebraska was selected as the host state. US Ecology then specified several stages in the development of the design: identifying basic requirements; conducting trade-off studies to determine the preferred method for meeting those basic requirements; developing detailed design criteria using the insights gained during the trade-off studies; completing preliminary designs to meet the detailed criteria; reviewing the design in conjunction with the characteristics of the site, once this is identified; and refining the final detailed design (US Ecology, 1988).

SITE SELECTION FOR THE PERMANENT DISPOSAL FACILITY

Initially, US Ecology developed a broad statement of objectives to direct the site-selection process. These objectives were designed to lead to the selection of three candidate sites that could protect public health and safety; provide a geotechnically and environmentally suitable location; be licensed in a timely manner after characterization; be located in an area that expressed interest in participating in the state screening process; and adhere to the requirements and deadlines of the federal Low-Level Radioactive Waste Policy Act, as amended (US Ecology, 1988).

By April 1988, US Ecology had identified and developed preliminary criteria for the several phases of the site-selection process (GAO, 1991). Five public meetings were held across the State of Nebraska from May 16 through May 23, 1988[55] (US Ecology, 1988). The final selection criteria are shown in Table 25.

In January 1993, the Nebraska Departments of Health and Environmental Control issued an intent to deny the license application indicating that the proposed site included wetlands. In response to the intent to deny, U.S. Ecology reconfigured the site, reducing the original plot of land from 340 to 120 acres, thereby omitting the wetland portion of the site.

Also in January 1993, Governor Nelson filed suit in federal court to block utilization of the site indicating that community consent, a state requirement for selecting the site, had not been obtained. On September 30, 1993, the suite was dismissed by the federal judge. Following dismissal of the federal lawsuit, the Nebraska Departments of Health and Environmental Control announced that they were going to withdraw the intent to deny the license.

---

[55] The public meetings were moderated by a representative from the Nebraska League of Women Voters. The meetings were designed to provide the public with information about the siting process and to obtain public input about the criteria. Attendance at these public meetings ranged from 15 to 120 persons (US Ecology, 1988).

Table 25. FINAL SITE SELECTION CRITERIA

1) Groundwater
    1.1) outside influence area of public water well;
    1.2) away from area with shallow fluctuating groundwater;
    1.3) away from major groundwater recharge zones;
    1.4) away from areas with complex hydrologic conditions;
2) Geology
    2.1) stable tectonic processes;
    2.2) two miles or more from active faults;
    2.3) avoid areas with sand and gravel deposits on the surface with shallow depth to ground;
    2.4) geologic hazards such as subsidence, mass wasting, slumping, liquefaction;
    2.5) geologic resources (e.g., mineable sand and gravel deposits, mineral concentrations and hydrocarbon deposits) that, if developed, could affect the safe operation of the facility;
3) Surface water
    3.1) outside 100-year flood plain;
    3.2) outside flood prone areas;
    3.3) away from water bodies (lakes, rivers, creeks, canals, and ponds) and wetlands;;
    3.4) away from areas with drainage problems (e.g., local ponding);
4) Land use
    4.1) outside boundaries of legally dedicated land;
5) Population and Urban Growth
    5.1) areas 15 or more miles away from population centers over 100,000;
    5.2) areas 2 or more miles away from population centers over 5,000
    5.3) two kilometers (1.2 miles) or more away from population centers;
6) Cultural resources
    6.1) outside national register sites;
7) Biological resources
    7.1) outside designated critical habitat for federal or state listed threatened or endangered species; and
8) Community compatibility
    8.1) formal invitation by host entity to evaluate the site.

Source: GAO, 1991.

In October 1993, the Board of the Natural Resources District that encompasses Boyd County indicated that they will file a lawsuit to prevent siting the facility.

The purpose of the initial screening process was to identify geographic areas within the State likely to include a suitable site. This phase of the process consisted of identifying suitable geographical areas; obtaining from counties and local communities formal expressions of interest in participating in the screening process; and identifying potential sites within the general areas expressing interest (GAO, 1991).

US Ecology relied on a subset of criteria to conduct an initial screening of the State. These criteria required that potential sites be located outside the 100-year flood plain; outside boundaries of legally dedicated land; in areas 15 or more miles from population centers over 100,000; and in areas two or more miles from population centers over 5,000 (GAO, 1991).

In June 1988, US Ecology solicited input from counties and communities to determine their willingness to participate in the state and regional screening process. Any county not expressing interest at this time was excluded from further consideration. By August 1988, 20 counties had formally expressed interest in the screening process (GAO, 1991). The village of Butte in Boyd County passed a resolution stating that the village would work cooperatively with the contractor to objectively evaluate suitable sites in the region and submitted a letter of interest to US Ecology.

Areas in interested counties remained under consideration if they were not excluded on the basis of a number of screening criteria.[56] The areas remaining after this screening were designated "potential areas" and were evaluated further to identify smaller areas within them. These smaller areas, ranging from 1/2 mile to 18 square miles, were designated as "potential siting areas" (GAO, 1991).

Three counties withdrew their expressions of interest during the county-level screening process, while an additional six counties withdrew their expressions of interest after the potential areas had been identified. In the

---

[56] These included being outside the influence area of public water wells; away from areas with shallow fluctuating groundwater; away from major groundwater recharge zones; away from areas with complex hydrologic conditions; in areas with stable tectonic processes; in areas two miles or more from active faults; away from areas with sand and gravel deposits on the surface with shallow depth to ground; away from areas with geologic hazards such as subsidence, mass wasting, slumping, and liquefaction; away from areas with geologic resources (e.g., mineable sand and gravel deposits, mineral concentrations and hydrocarbon deposits); outside flood prone areas; away from water bodies (lakes, rivers, creeks, canals, and ponds) and wetlands; away from areas with drainage problems (e.g. local ponding); outside boundaries of legally dedicated land; areas two kilometers (1.2 miles) or more away from populations centers; outside national register sites; outside designated critical habitat for federal or state listed threatened or endangered species (GAO, 1991).

remaining 11 counties, there were 111 potential siting areas comprising 522 square miles (GAO, 1991).

In October 1988, US Ecology began screening the siting areas to locate available land at least 320 acres in size, measuring 1/2 mile by one mile, and owned by someone agreeing to grant an option to purchase the land at a future date, renew that option periodically, and grant access for site characterization work. Land agents were then used to locate tracts of land within the siting areas and sign purchase option contracts with receptive landowners (GAO, 1991).

In November 1988, the Citizens' Advisory Committee held a public meeting. At this meeting, the CAC participated in a two-part exercise to help test the relative importance of individual criterion. Each member of the Committee applied the criteria to 27 unidentified representative potential siting areas selected by US Ecology. The committee members generally favored 13 potential siting areas, most of which were in Boyd and Nemaha counties. After applying the criteria to the unidentified siting areas, the CAC went through the same exercise with the siting areas revealed. The CAC made few changes in their recommendations (League of Women Voters of Nebraska, 1989). According to US Ecology, this exercise confirmed the importance of groundwater and other geologic factors in selecting potential sites (GAO, 1991).

Following site visits by US Ecology's technical screening and siting personnel, potential siting areas in Boyd, Clay, Nemaha, and Nuckolls counties were viewed as having the greatest likelihood of being licensable. Previously, US Ecology and the Citizens' Advisory Committee had established a condition that the three candidate sites be selected from different groundwater geology regions. Since Clay and Nuckolls counties were in the same region, Nuckolls was selected because of more favorable groundwater characteristics (GAO, 1991).

US Ecology completed the selection of the three candidate sites in December 1988. Subsequently, Governor Orr was notified of the selection (US Ecology, 1991).

On December 6, 1988, the Butte Village Board of Trustees reaffirmed its support of the disposal facility. In a letter dated December 22, 1988, the Boyd County Board of Supervisors requested that US Ecology agree to several conditions related to community support, public health and safety, and economic compensation and reimbursement. The County Board also requested that the facility be an above-ground structure; a study of economic impacts be conducted; and $1 million for public improvement projects be guaranteed to the County. On December 30, 1988, US Ecology responded to the County Board, stating that a number of the requested conditions were already incorporated in the siting process and facility design and that the contractor did not have the authority to agree to all the remaining conditions (GAO, 1991).

The public announcement of the candidate site selection was made in January 1989 when landowner negotiations were completed and options to purchase agreements were signed. The timing of the announcement met the state's legislated target date to select three candidate sites. The Boyd County Board of Supervisors withdrew its support for the disposal facility a few days before the announcement of the three candidate sites, however, because US Ecology was unable or unwilling to meet the conditions imposed by the Board. US Ecology was not informed of the withdrawal of support until the day of the announcement (GAO, 1991).

After the three candidate sites were announced, and in accordance with Nebraska law, LMCs were formed to provide a means for local input and involvement in the siting process. While the Legislature envisioned that the LMCs would be comprised of interested local citizens of various backgrounds, membership was not legislatively restricted based on stated positions regarding the facility. Representatives of both opposition and support groups were appointed to the LMCs.

With the exception of the Nemaha County LMC, these committees soon became battlegrounds. In Nuckolls and Boyd Counties, many opponents of the facility or siting process attended LMC meetings and created disturbances. In February 1989, a federal EPA employee was appointed to a seat on the Nuckolls County Committee by a community opposed to the project. The appointment was successfully challenged in court, but by then the Nuckolls County site had been removed from consideration (*Lincoln Journal*, February 25, 1989; *Hastings Tribune*, March 31, 1989).

US Ecology hired a contractor, Bechtel International, Inc., to help conduct site characterization. In January 1989, they began the effort to characterize each candidate site. Site characterization included drilling bore holes and well holes to assess the geology and groundwater, assessing surface water, and mapping water drainage patterns. The LMCs hired technical experts to monitor the characterization work (GAO, 1991).

By mid-December 1989, site characterization had been completed. According to US Ecology, all three sites appeared to meet or exceed federal and state licensing requirements. The site in Boyd County was attractive because its relatively simple geology allows geological characterization with a high degree of reliability; 500 to 1,000 feet of impermeable shale bedrock provides a natural barrier to the aquifer system; and US Ecology determined that the Boyd County site was technically superior to the two other sites (US Ecology, 1990).

Two local citizens' groups were formed in Boyd County to address the LLRW disposal facility issue. The People for Progress, formed before Boyd County was the preferred site, supports the idea of locating the facility near

Butte. According to press reports, they support the facility because they believe it can be built and operated safely and that it will provide more jobs, lower taxes, and better educational opportunities. An opposition group, Save Boyd County, was created when Boyd County became one of the three potential sites. According to press reports, this group believes that Nebraskans' financial resources, environment, and way of life will be jeopardized if the LLRW disposal facility is built anywhere in Nebraska (*Norfolk Daily News*, September 19, 1990).

On July 27, 1990, US Ecology formally submitted to the State of Nebraska an application for a license to construct and operate a LLRW disposal facility 2.5 miles west of Butte in Boyd County.[57] The license application includes an Environmental Report and a Safety Analysis Report presenting information from studies of the site and the surrounding area, details of the facility design, descriptions of operational procedures, and evaluations of potential impacts of the facility on the area (US Ecology, 1990).

According to the Environmental Report, several short-term and long-term impacts at the Butte site were considered.[58] The Safety Analysis Report addressed the ability of the facility to meet or exceed technical safety requirements. The performance of the proposed facility was assessed for operational, closure, and post-closure phases in terms of potential public health and environmental risks. Several normal operating situations and "worst-case" scenarios were examined to determine potential off-site doses to the general public. According to the report, the calculated dose in each case was well within regulatory limits of 25 millirems (US Ecology, 1990).

In October 1990, DEC and DOH submitted a Notice of Deficiency to US Ecology highlighting information that was missing from the license application. US Ecology supplied the information missing from the license application and, in December 1991, DEC and DOH certified that the license application was complete. Early 1992, the State issued its second round of comments on the application. US Ecology is currently working on the responses to the State's technical questions.

During the November 1990 general election, a candidate opposed to the LLRW facility replaced an incumbent on the Lower Niobrara Natural Resources Defense (LNNRD) board. In January 1991, the LNNRD replaced

---

[57] According to the most recent census, 2,800 people live in Boyd County.

[58] These included the influx of revenues into the community; an increase in non-farm employment in the area; an increase in demand for temporary housing during construction phases; intermittent construction noise; increases in existing background level for some air pollutants during construction; potential sediment loading in storm water runoff during construction; the removal of 320 acres from private use (227 acres are prime agricultural farmland); disruption of 120 acres of vegetation/habitat during construction and closure; potential for radiological contamination; and reduced employment at the site after closure (US Ecology, 1990).

its two representatives on the LMC. While the previous representatives supported the facility, their replacements were opposed to it (*Omaha World Herald*, January 9, 1991).

The LMC canceled or postponed several monthly meetings in 1990 because of a lack of quorums and operating funds. The Committee has also closed their office. Opponents criticized the LMC chairman, a project supporter, alleging that he was responsible for a lack of activity on the part of the LMC. The chairman and other LMC members responded that a legal dispute mounted by waste opponents in another county had blocked LMC funding from US Ecology.

In February 1991, the new LNNRD appointees and two other LMC members opposed to the project called for a meeting of the LMC. The chairman responded by challenging the validity of the LNNRD appointments (*Omaha World Herald*, February 23, 1991).

In March 1991, another dispute arose regarding LMC membership. When the Village of Butte appointed a local dairy farmer to fill a vacant position, two communities opposed to the facility disputed Butte's authority to make the appointment. Butte officials replied that the State agreed that the appointment was theirs to make. During a subsequent LMC meeting, one of the State's leading facility opponents stated that she was there to represent the two communities opposed to the project and attempted to take a seat on the LMC. The LMC refused to recognize her and voted to adjourn. The State Attorney General issued an opinion that the woman was not eligible to serve in the position on the LMC because she was not a resident of Butte (*Holt County Independent*, March 1, 1991).

On August 14, 1991, Governor Nelson announced that he had replaced the LMC member serving as chairman. The Governor argued that the new member was neutral on the issue. Many supporters disagreed, stating that the appointee had signed a petition in 1988 opposing the site and written letters to a newspaper opposing the facility. The Governor's decision was viewed as a positive step by facility opponents. A member representing the community of Butte resigned. On August 19, 1991, the LMC met and selected a leading project opponent as its new chairman (*Lincoln Journal*, August 13, 1991).

During 1990, a lawsuit was filed by Concerned Citizens of Nebraska against US Ecology and the Compact. The suit alleged violations of the United States Constitution and the federal Low-Level Radioactive Waste Policy Amendments Act. All claims against US Ecology and the Compact were dismissed by the Federal Court (GAO, 1991).

During 1990 and 1991, several incidents of harassment and violence occurred between supporters and opponents in Boyd County. For example, shots were fired into the home of the chairman of the LMC, protest signs were repeatedly torn down, and two teenagers were threatened by armed men when attempting to remove a protest sign (*Norfolk Daily News*, April 3, 1990; *Omaha World Herald*, July 31, 1991).

McCulley Township, located in Boyd County, where the LLRW disposal facility would be situated, is a rural government entity with approximately 80 registered voters. Voters in the Township passed regulations prohibiting the storage of radioactive materials within the Town on October 28, 1991. The vote also authorized the Township Board to take legal steps to enforce the ban. Both votes passed by a 43-4 margin. This vote followed a vote in August 1991 banning the disposal of injurious substances in the Township. Attorneys for the Compact Commission indicated that actions of McCulley Township would be ineffectual in blocking the facility because those actions are superseded by State and federal laws. Governor Nelson indicated that a court of the Nebraska Legislature would have to determine whether the regulations passed by the Township would prevent a waste facility from being built there (*Omaha World Herald*, October 30, 1991).

In October 1991, the Compact Commission held their first Boyd County meeting in Butte. While 50 supporters welcomed the facility as a means of creating jobs, approximately 300 opponents protested the meeting and staged a mock funeral and burial of State, compact, and local officials. During the meeting, the meeting hall was vandalized. Because of the protests, no Compact business could be discussed, and the day-long meeting focused solely on public comments. The comments voiced during the day suggested to the Commission that Butte, at 2.5 miles, the closest municipality to the proposed site, tended to favor the site while Spencer, ten miles away, staunchly opposed the proposed facility. The Compact Commissioners cancelled a planned tour of the proposed site due to personal safety concerns (*Omaha World Herald*, October 23, 1991).

On December 18, 1991, a controversy arose when a US Ecology representative attended a Boyd County Local Monitoring Committee meeting in Naper with six bodyguards. The project manager at US Ecology indicated that he did not feel comfortable attending the meeting without some form of personal protection. Several people indicated that guards stationed at the door told people that they were carrying weapons but did not show them. Following the incident, a local State senator requested that the State Patrol and Governor's office investigate the allegation. US Ecology representatives stated that they welcomed the investigation but argued that any probe into the incident should also determine whether anyone in the audience was armed. The guards denied the allegations that they were armed at the Naper meeting, and the subsequent investigation was inconclusive (*Lincoln Journal*, December 20, 1991; *Omaha World Herald*, December 24, 1991; January 16, 1992).[59]

---

[59] According to US Ecology, company representatives have subsequently attended LMC meetings with reduced security (US Ecology, 1991).

In August 1992, public opposition escalated to a new level when two State employees were prevented from getting out of their car at the Local Monitoring Committee's meeting. The incident was followed by theft and damage to scientific equipment at the site estimated at $7,500 (*The Radioactive Exchange*, Vol. 11, No. 16, 1992).

In January 1993, the Nebraska DEC and Department of Health announced their intent to deny the license application because the proposed site included wetlands. In response, U.S. Ecology reconfigured the site, reducing the original plot of land from 340 to 120 acres, thereby omitting the wetland portion of the site.

Also in January 1993, Governor Nelson filed suit in federal court to block use of the site, arguing that community consent, a state requirement for seleting the site, had not been obtained. On September 30, 1993, the suit was dismissed by the federal judge. Subsequently, the Nebraska DEC and Department of Health reversed their announced intention to deny the license.

METHOD SELECTION FOR THE PERMANENT DISPOSAL FACILITY

Prior to selection of a host state, US Ecology indicated that it intended to build an above-grade facility. US Ecology and its subcontractor, Bechtel, proposed using an above-grade facility because of public preference, regulatory requirements, and technical considerations. According to the contractor, the consensus in numerous public meetings throughout the compact states was that an above-grade disposal system would allow better visual monitoring of system performance and greater flexibility in terms of retrieving of the waste during the operational period (US Ecology, 1988).

Several design and planning requirements were specified in Nebraska legislative bills LB 426 and LB 761 and became part of the Nebraskan LLRW Disposal Act. Traditional shallow land burial, as used prior to 1979, was forbidden, and construction of disposal cells above grade was required. The law required that the license applicant present a design for above-ground disposal or other technology that contains at least one engineered, artificial barrier to isolate waste from the surrounding environment. The law also required that the facility design include a plan for retrievability and removal of all waste (Nebraska, State of, LB 426, 1987).

The Citizens' Advisory Committee reviewed and commented on the facility design. These comments, as well as other public input, resulted in the facility design being changed from an open-topped structure to a totally enclosed structure to accommodate concerns about rainfall and tornados (League of Women Voters of Nebraska, 1989; US Ecology, 1991).

COMPENSATION AND INCENTIVES

Several of the conditions established by Governor Orr when Nebraska accepted the responsibility of hosting the disposal facility focused on compensation and incentives. Specifically, reasonable costs incurred by local monitoring committees had to be reimbursed; compensation to the State and host community had to be guaranteed at reasonable levels for the operating life of the facility; the State had to be reimbursed for all costs pertaining to regulating, licensing, and planning the facility; compensation to local communities involved in the siting process had to be guaranteed; and property values in the area surrounding the site had to be guaranteed (Governor Orr, undated).

In 1988, amendments to the Nebraska LLRW Disposal Act (Nebraska, State of, LB 1092, 1988) established the Community Improvements Cash Fund. This fund provides compensation at acceptable levels to the local communities who become active participants in the site-selection process and to the State and host community when the final site is selected. Legislation passed in 1989 (Nebraska, State of, LB 761, 1989), however, removed the State as a recipient of these funds during and after completion of the site-selection process. Monies for the Community Improvements Cash Fund comes through the Compact Commission from the non-sited states. The monies are paid annually to the DEC, which administers the money (Nebraska, State of, 1990/b; US Ecology, 1991).

During the Site Characterization phase, $300,000 was distributed equally among each of the three candidate sites in Nuckolls, Nemaha, and Boyd counties. Once the Boyd County site had been identified as the preferred site, that area began to receive $300,000 annually. Fifty percent of the Community Improvements Cash Fund is divided between the incorporated municipalities within ten kilometers of the site. Distribution of this money is based on the ratio of population of each incorporated municipality to the total population of all eligible municipalities as determined by the latest federal census.[60] The remaining 50 percent of the money is forwarded to the County Treasurer of the affected county, to be distributed to each political subdivision that levies property taxes on the facility site. The money is distributed on the basis of the ratio of the total amount of property taxes levied by each political subdivision to the total amount of property taxes levied by all such political subdivisions on such property (Nebraska, State of, 1990/b; US Ecology, 1991).

Once the facility becomes operational, Nebraska law requires that $2 million a year be provided to the State by the developer for distribution in the Boyd County area according to the same distributional formula. Nebraska

---

[60] The eligible communities are Butte and Monowi.

has placed few restrictions on use of money from this fund. The funds are to be used for public purposes described as the "promotion of the public health, safety, welfare, and environment of the inhabitants" (Nebraska, State of, 1990/b). In Boyd County, the money has been used to help build a new fire hall and a shop and maintenance building in Butte, for street maintenance in several communities, and to improve services at the County's only hospital, located in Lynch (US Ecology, 1991).

US Ecology has worked with People for Progress, a Boyd County organization supporting development of the facility, to identify and provide grants to worthwhile projects in need of financial assistance. Through the first seven months of 1991, US Ecology donated approximately $39,273 to 17 different projects or groups (*Norfolk Daily News*, August 3, 1991).

The 1988 amendments to the Nebraska LLRW Disposal Act (Nebraska, State of, LB 1092, 1988) also created and provided these committees with a source of operating funds. Two separate funds were created to handle disbursement of the money to the LMCs. The first, the Local Site Selection Cash Fund, was used by the State to distribute a one-time $300,000 dollar grant to the LMCs in the candidate site areas. The second fund, the Local Monitoring Committee Cash Fund, was put into operation once the preferred site was selected. Through this fund, the Boyd County Local Monitoring Committee is to receive $100,000 annually throughout the licensing, construction, operation, closure and custodial care phases of the facility. State law requires US Ecology to provide monies for these funds (Nebraska, State of, 1990/a; US Ecology, 1991).

After the license is issued, the LMC can use these funds to hire or contract with a qualified disposal facility inspector as determined by the DEC. The LMC may hire clerical staff and purchase office supplies but may not hire professional and technical staff other than an inspector. However, the LMC can contract for professional and technical services (Nebraska, State of, 1990/a).

The amendments to the Nebraska LLRW Disposal Act authorize compensation for loss of property value near the facility. Any property owner within three miles of the facility on the date a license for the facility is granted who believes his or her property value has declined may apply for compensation. An application for compensation must be filed within five years of the date that the facility begins operation. Upon application by the property owner, the county board shall hold a hearing to determine whether a loss of property value has occurred. If the property owner establishes, by a preponderance of the evidence, that his or her property has suffered loss, the developer must provide compensation for the amount of the loss (Nebraska, State of, 1990/a).

DEC is required to offer landowners directly adjacent to the facility's boundary an annual well and surface water sampling and analysis of any domestic water supply at no cost to the landowner. The Department is also required to offer landowners directly adjacent to the facility's boundary the opportunity for testing of agricultural products produced on the property for contamination by radioactivity (Nebraska, State of, 1990/b).

In addition, the law requires that the developer, US Ecology, provide training for the first responding fire, police, and ambulance services to handle emergency events at the facility and support for affecteð county emergency management planning, training, and central dispatch facilities. The developer must also provide to the appropriate local political subdivision any equipment that is necessary to provide emergency response capabilities due to the location and operation of the facility (Nebraska, State of, 1990/b).

INTERIM WASTE MANAGEMENT

Because virtually all of the LLRW produced in Nebraska is generated by the State-owned nuclear power plants, interim waste management has not emerged as an important issue. After access to out-of-State disposal facilities is lost, power plants will store waste on-site until a facility is available.

In early 1993, all states in the Central Interstate Compact were denied access to the Barnwell disposal facility. However, in October 1993, the Southeast Compact voted to renew access of the Central Interstate Compact states to the Barnwell facility through 1994.

ANALYZING THE NEBRASKA CASE HISTORY FROM THE PERSPECTIVE OF THE COMPETING VALUES APPROACH

As was true in several of the other states that we studied, the Nebraska siting process was more goal-oriented and efficient in its early stages than it was later on. In 1983, Nebraska joined the Central Interstate Compact, and, in 1987, was selected as the host state. U.S. Ecology was selected as the contractor and initiated the site- and method-selection processes, which resulted in the submission and completion of a license application in less than five years, a relatively short time in comparison to other states' experiences. In 1990, however, a new Governor, who opposed building a LLRW disposal facility in the State, was elected, and facility construction has been blocked. It now appears likely that the process will ultimately prove much less efficient from the rational perspective than initially appeared likely.

From the empirical perspective, which emphasizes the use of data-based processes and accountability in decision making, the Nebraska process was comparatively strong, even though, unlike most states that were strong from this perspective, the Nebraska process did not make identification of the technically optimal site an explicit goal. Rather, U.S. Ecology emphasized the desirability of selecting a site that could be licensed in a timely fashion and actively sought communities to volunteer as potential host sites. Nonetheless, the process used a great deal of data in the site-selection process. Twenty screening criteria were applied in sequential fashion until three potential sites were identified. U.S. Ecology then selected the site in Boyd County on the grounds that site characterization data showed it to be technically superior to the other two sites. Although the process appears to have striven for accountability, inasmuch as extensive documentation is available, the justification for selecting the Boyd County site over the other two potential sites has been questioned by some stakeholders.

From the consensual perspective, which emphasizes participatory processes and supportable decisions, the Nebraska case seems stronger on process than it proved in results. Both the State of Nebraska and U.S. Ecology placed greater emphasis on involving the public in the decision making process than was typically the case. Before joining the Central Interstate Compact, Nebraska created a Citizen's Advisory Board on LLRW. After Nebraska was selected as the host state, the State legislature passed a law requiring community consent to the siting of a facility (although the standard for assessing community consent was not clearly defined and subsequently led to debate).

At the beginning of the site selection process, U.S. Ecology asked counties and communities to indicate their interest in participating in the site screening process; any county not expressing interest at this time was removed from further consideration. The Citizen's Advisory Committee participated in a process to determine the relative importance of the siting criteria. After the three candidate sites were announced, Local Monitoring Committees (LMC) were formed in each of the three counties. After identification as the potential host site, public opposition to siting the facility began to emerge in Boyd County. Eventually, opponents to the facility outnumbered its supporters on the LMC.

Despite the emphasis on participation -- soliciting volunteer communities and encouraging public involvement at all stages of the process -- the Nebraska process did not result in a decision that enjoined universal support. Although the community closest to the site has generally supported the decision, a community ten miles away continues to strongly oppose the facility.

Finally, the political perspective emphasizes adaptable decision making processes designed to achieve decisions that are perceived to be legitimate. Because the site- and method-selection processes in Nebraska were not prescribed by the State, but rather were developed by the contractor, decision making processes were comparatively flexible. Site and method selection plans did not have to be released to the public and, therefore, U.S. Ecology was not subject to criticism that it failed to follow some previously specified plan.

The legitimacy of the process was subjected to severe scrutiny when the former director of the Compact Commission, convicted of embezzling Compact funds, alleged that the Boyd County site was selected for political reasons. Although there remains some dispute and confusion over these charges, a U.S. General Accounting Office inquiry concluded that the selection of the Boyd site was proper. At least one credible, outside evaluator, then, has judged the process to be essentially legitimate.

TEXAS

BACKGROUND

In 1979, approximately 543 cubic meters (410 curies) of LLRW was generated in Texas. By 1990, these figures changed to 261 cubic meters and 1,600 curies (Table 26). In 1990, 15 percent of the waste produced came from governmental and military organizations, 23 percent was generated by nuclear power plants, and 62 percent came from institutional and industrial sources (DOE, 1980, 1981, 1982, 1983, 1984/a, 1985, 1986/a, 1987, 1988, 1989, 1990, 1991).

Table 26. TOTAL LOW-LEVEL WASTE GENERATED IN TEXAS AND DISPOSED OF AT COMMERCIAL DISPOSAL SITES BY YEAR

	Volume (cubic meters)	Radioactivity (curies)	Volume Percentage by Source		
			Institutional and Industrial	Power Reactor	Government and Military
1979	543	410	100%	0%	0%
1980	2,131	717	99%	0%	1%
1981	2,477	1,012	99%	0%	1%
1982	1,856	257	99%	0%	1%
1983	1,628	1,547	99%	0%	1%
1984	364	263	96%	0%	4%
1985	299	4,841	94%	0%	6%
1986	124	3,094	35%	0%	65%
1987	1,959	7,027	22%	0%	78%
1988	342	1,853	72%	0%	28%
1989	627	11,357	80%	9%	11%
1990	261	1,650	62%	23%	15%

Source: DOE, 1980 - 1991.

A chronological summary of key events related to the siting of a LLRW
disposal facility is given in Table 27.

Table 27. CHRONOLOGICAL SUMMARY OF KEY EVENTS IN TEXAS

Date	Event
1981	Texas Legislature passed the Texas LLRW Disposal Authority Act. The Texas LLRW Disposal Authority (Authority) was established.
1982	Ebasco Services, Inc., was selected to develop facility design.
1983	Texas Department of Health promulgated TRCR Part 45 regulations on siting requirements.
1985	The Authority selected a preferred site in McMullen County. The Texas LLRW Disposal Authority Act was amended to prohibit siting the facility within 20 miles upstream from a reservoir; entitle the Authority to purchase state-owned land; and ban shallow land burial.
1987	The Authority selected a site near Fort Hancock. El Paso County filed lawsuit against the Authority.
1989	The Authority announced the Fort Hancock site.
1991	The Court voided the site-selection process and found the Fort Hancock site unsuitable. The Texas LLRW Disposal Authority Act was amended to define a geographic area in eastern Hudspeth County in which the site must be located; grant the Authority the right of eminent domain; and authorize Texas to enter into an interstate compact. The Authority selected the Fasken Ranch site. Generators developed detailed plans for storing or otherwise managing their waste until the Texas facility is operational.
1992	The Authority submitted the license application for a design including below-ground modular canisters. The Authority purchased the Fasken Ranch property. Negotiations were initiated with Maine and Vermont about the possibility of an interstate compact.

OVERALL APPROACH

The rapidly increasing volume of LLRW generated in Texas, combined with
emerging problems at the three commercial disposal sites in Nevada, South
Carolina, and Washington signaled the start of a potential crisis in LLRW
disposal in 1979. In the 1979 session of the Texas State Legislature, two
bills were introduced which were intended to regulate both uranium mill
tailings[61] and LLRW disposal by giving the Department of Health control

---

[61] The primary purpose of the legislation was to meet the mandates of the Uranium Mill
Tailings Control Act (UMTA). The proposed legislation, however, was much broader in
scope than required by UMTA.

over all low-level radioactive material and disposal facilities. Opponents of the legislation proposed waste reduction and continued reliance on out-of-state disposal. After much debate, both bills died in session (Blackburn & Alvarado, 1985; Mathews, 1985).

The 1980 federal LLRW Policy Act made it clear that out-of-state disposal was not a tenable long-term solution to the Texas LLRW problem. There was little support for authorizing the State's participation in a compact, however, because of the widespread belief that Texas, one of the largest generators in the United States, would become the host state in any such compact (Mathews, 1985).

In 1981, the Legislature made a new attempt to act on LLRW management. A bill introduced in both houses proposed establishing a LLRW facility exclusively for waste generated in Texas. Known as the Texas Low-Level Radioactive Waste Disposal Authority Act (Texas, State of, Ch. 273, 1981), this law was passed by the Legislature and signed by the Governor in June 1981.

The law created the Texas Low-Level Radioactive Waste Disposal Authority (Authority), to be governed by an independent board of directors. It instructed the Authority to select a site and a technology for a disposal facility, and to construct, operate, and finance it. After screening the State, the board was to select two or more sites for further analysis. The law dictates that the board select "the site that appears from the studies to be the most suitable site" from those which had been subjected to further analysis (Section 3.07). The facility was to be regulated and licensed by the Department of Health, Bureau of Radiation Control.

The Act required that all expenses of the Authority, including the costs of decommissioning, closure, maintenance, and impact assistance to local communities, be covered by waste generators. The law prescribed that the Authority acquire the fee-simple title (i.e., ownership of surface and mineral estates) to all land and property that is part of the disposal site. The mandate did not provide the Authority with the power of eminent domain (i.e., the right to condemn land for the site).

When, after a three-year effort, the Authority planned to announce a preferred site in McMullen County, South Texas, in January 1985, South Texas legislators intervened. A bill was introduced to extend the responsibilities of the Authority from waste disposal to waste management, which could include waste processing and storage. A second bill directed the Authority to investigate alternative technologies and additional sites and delay the siting. Proponents of this bill argued that Texas was moving too fast with siting and if Texas opened a site before the other states, it would make it a prime candidate for becoming the nation's fourth disposal facility (Mathews, 1985).

In 1985, amendments to the Texas LLRW Disposal Authority Act were passed by the Legislature (Texas, State of, Ch. 692, 1985). The amendments prohibited siting the facility within 20 miles upstream from a reservoir constructed or planned for construction by the U.S. Bureau of Reclamation or the U.S. Army Corps of Engineers.

This intervention excluded not only the proposed site, but a large part of practically all of South Texas from consideration, and directed the Authority to concentrate on State-owned lands in West Texas. The problem was, however, that according to Texas law, the General Land Office had unilateral authority over State-owned land. In order to utilize State-owned land, another provision had to be added to the amendments which entitled the Authority to purchase State school land from the General Land Office and universities (Texas, State of, Ch. 692, 1985).

The 1985 amendments banned shallow land burial and instructed the Authority to investigate alternative technologies. The law also directed the Department of Health not to issue a license prior to July 1987. This provision was intended to enable the Legislature to intervene if necessary during its 1987 session.

The search for new sites concentrated on State-owned lands in West Texas. In August 1986, the Authority named two sites in Hudspeth County as sites for further consideration. In February 1987, the Authority was ready to designate a site located ten miles northeast of Fort Hancock (Hudspeth County) and about 45 miles east of El Paso (El Paso County).

In February 1987, El Paso County filed a lawsuit against the Authority contending that the disposal facility would have an adverse impact on El Paso County; that the site was too close to a major reservoir; and that the Authority did not select the "most suitable" site, as the law required (Fowler, 1989). Following this, a State district court granted a temporary injunction halting site selection. In August 1987, the 8th Court of Appeals reversed the temporary injunction, and the Authority resumed its site selection work in late 1987.

In November 1989, the Authority announced the selection of the Fort Hancock site, and the lawsuit was reinstated by El Paso County. El Paso County officials immediately obtained another temporary injunction preventing further work. In January 1991, the State district court in Hudspeth County voided the site-selection process and found the Fort Hancock site unsuitable. The court judgment claimed there was a potential for earthquakes and a danger of water contamination in the area. The court issued a permanent injunction stopping the siting process. The Authority appealed the case in April 1991, but abandoned it when the Legislature further amended the law.

In a letter of March 15, 1991, the Governor urged the Authority to abandon the Fort Hancock site and find another site in Hudspeth County because of the intensity of the opposition and the judge's ruling. Thereafter, negotiations were initiated among representatives of the Authority, and El Paso and Hudspeth counties. In May 1991, the Legislature passed new amendments to the LLRW Authority Act (Texas, State of, Ch. 402, 1991). The amendments defined a geographic area of 400 square miles in eastern Hudspeth County, about 100 miles from El Paso, in which the site must be located. Most of the designated land is privately owned. To facilitate the purchase of land, the amendments granted the Authority the right of eminent domain.

The May 1991 amendments authorized Texas to enter into a compact with one or more states for disposal of their waste. The law limited the volume of waste imported from other states to 20 percent of the waste generated in Texas. Under the new law, the states wishing to compact with Texas have to contribute to the siting expenses and construction of the facility. A certain part of this contribution will be granted to the host county.

In response to the above legislation, the Governor's Office implemented a compact task force. During 1992, Maine and Vermont contacted the task force regarding the possibility of compacting with Texas.

INSTITUTIONAL STRUCTURE

Texas does not have a single agency for environmental protection; therefore, multiple state agencies have to be involved in regulating any industrial facility. With respect to LLRW disposal, the Texas Water Commission regulates facilities for water quality, the Texas Air Control Board regulates air quality, the Texas Department of Parks and Wildlife regulates endangered plants and animals, and the Bureau of Economic Geology of the University of Texas acts as the state geology office.

Texas has been an agreement state since 1963, and the Texas Department of Health is the designated agency regulating the management of radioactive material. The LLRW Disposal Authority Act (Texas, State of, Ch. 273, 1981) created the Texas Low-Level Radioactive Waste Disposal Authority and authorized it to select a site and to construct, operate, maintain, decommission, close, and finance a disposal facility in Texas. Under the law, the Authority could choose to contract with a private firm for day-to-day operation, with management control retained by the Authority.

The Authority was formed in 1981. According to the terms of the LLRW Disposal Authority Act, its board of directors includes a medical doctor, a health physicist, an attorney, a geologist, and two members of the general public. The act specified that after a site had been selected, a resident of the host county was to be appointed to the board.

The 1987 amendments (Texas, State of, Ch. 76, 1987) to the LLRW Disposal Authority Act instructed the Board to create a local Citizens' Advisory Committee to oversee the facility. The Committee was to be established within 30 days after construction of the facility began. It is authorized to conduct independent monitoring and to make recommendations to the Board concerning operation of the facility.

In 1991, a law changing the institutional structure of environmental regulation was enacted (Texas, State of, Ch. 583, 1991). The law changed the Texas Water Commission into a regulatory agency for overall environmental conservation. According to the law, the license application for the LLRW facility will also be reviewed by the Texas Water Commission.

DESIGN OF THE SITE- AND METHOD-SELECTION PROCESSES

The Texas LLRW Disposal Authority Act of 1981 directed the Authority to identify two or more potential disposal sites on the basis of technical criteria and to select a "most suitable" site from among a small group of identified sites, considering additional socioeconomic criteria. The law of 1981 did not include instructions about the method selection process.

The law directed the Authority to invite public officials and other members of local public entities to participate in the site-selection process and to hold a public hearing in the host county at the end of the process. The statute did not prescribe any other form of public participation.

There was a general assumption that shallow land burial would be applied as the disposal technology. In February 1983, the Authority contracted with Dames & Moore to conduct the site-selection study. The contractor proposed a three-phase study that included screening the entire State to determine preferred siting areas; identifying two to five potential sites for preliminary characterization; and designating and characterizing two or more preferred sites.

The Authority, recognizing the need for public participation in the site-selection process, created the Citizens' Advisory Panel in 1983. The Panel included six members with varying backgrounds representing all areas of the State. The Authority staff met regularly with the panel to receive guidance from them (TLLRWDA, 1984/e). The Panel was instrumental in defining and weighting site-selection criteria.

In 1984, the Authority tried the Keystone Process, a conflict resolution method based primarily on soliciting early public participation (Keystone Center, 1982). In two of the counties (Dimmit and Hudspeth) with sites identified for preliminary characterization, review committees were organized to identify important issues and to encourage a dialogue between the Authority and the affected public. The Authority planned to establish similar dialogue groups in other potential areas and sites (TLLRWDA, 1985), although these plans were never implemented.

The Authority also planned a series of public meetings and public opinion surveys. In 1985, the Authority hired the Texas Advisory Commission on Intergovernmental Relations to suggest the most efficient way to involve local government officials in the siting process. The Commission investigated various types of involvement and recommended mitigation, negotiation, and incentives as important elements of public participation. The Commission said that local government officials should have a chance for input concerning the selection of a site operator (TLLRWDA, 1985).

The 1985 amendments to the LLRW Disposal Authority Act (Texas, State of, Ch. 692, 1985) banned shallow land burial and directed the Authority to investigate alternative disposal technologies. The amendments also instructed the Authority to start the selection process again by focusing on State-owned land. At this stage, the Authority planned to conduct site and method selection in parallel.

Site selection was planned to include three phases (Blackburn & Alvarado, 1985): screening all State-owned lands to identify potential siting areas; identifying preferred sites for preliminary characterization; and identifying a primary site for detailed characterization.

Method selection was planned to include the following phases (TLLRWDA, 1987): reviewing available disposal technologies; identifying disposal technologies for conceptual design; developing and analyzing conceptual designs of disposal facilities; and selecting a disposal technology.

The site- and method-selection processes would be integrated during the phase of conceptual design, when the characteristics of the preferred sites and methods would be considered. All the precluding activities were to be performed by the Authority and its contractors.

After the Fort Hancock site was selected as the primary site in 1987, the Authority increased its public outreach activities. In 1988, it opened a local office in Fort Hancock, to provide information to local citizens.[62]

Representatives of El Paso County and Hudspeth County were actively involved in the legal fights against the site-selection process. Their efforts resulted in amendments to the law in 1987, defining a local assistance

---

[62] The activity of the office included contacts with officials of Hudspeth County and El Paso County (the counties closest to the site), and other elected officials; meetings with residents of Fort Hancock, Sierra Blanca, and Dell City (the municipalities closest to the selected site), and presentations and dissemination of education materials; organized tours to the Beatty, Nevada, site and the DOE Nevada Test Site; contacts with the media; and contacts with representatives of the Mexican government.

package for the local community, and in 1991, moving the siting area about 30 miles southeast and granting additional aids to the local community.

SITE SELECTION FOR THE PERMANENT DISPOSAL FACILITY

The LLRW Disposal Authority Act directed the Authority to consider ten site-selection criteria (Texas, State of, Ch. 273, 1981).[63] In addition, the law required the Authority to take into consideration the costs of development, operation, maintenance, decommissioning, and extended care, as well as socioeconomic, environmental, and public health impacts. In 1982, the Texas Department of Health drafted regulations on "Licensing Requirements for Near-Surface Land Disposal of Radioactive Waste," which were finalized and promulgated in 1983 (TDOH, Part 45, 1983). The siting requirements are summarized in Table 28.

In 1982, the Authority developed its own guidelines (TLLRWDA, 1982) which took both the State and federal regulations into consideration. The guidelines are displayed in Table 29.

In 1983, the Citizens' Advisory Panel was invited to participate in the siting process. The Panel said that the most important siting issue should be minimizing the interaction of LLRW with groundwater or surface water. Other siting issues of concern noted by the Panel included transportation and geology.

In 1983, Dames & Moore conducted a Statewide screening[64] which excluded a number of unsuitable areas.[65] Dames & Moore identified 15 potential siting areas in 105 counties. In the next phase of site selection, the 15 areas were assessed with regard to a number of key criteria (Blackburn &

---

[63] These included: geology; surface characteristics; other aspects of transportation and access; meteorology; population density; surface and subsurface hygrology; flora and fauna; current land use; proximity to sources of LLRW; and criteria established by the agency for site selection.

[64] This was a discretionary exclusion at a certain scale and could be reexamined at a larger scale with closer site-specific examination.

[65] These included: geologically less suitable areas; major and minor aquifer recharge zones; areas of major mineral deposits; coastal high-hazard areas; national and state parks, forests, wildlife management areas, and refuges; and major populated areas and areas with high population density or potential future growth.

Table 28. SITING REQUIREMENTS

1. The disposal site shall be capable of being characterized, modeled, analyzed, and monitored.
2. Within the region where the facility is to be located, a disposal site should be selected so that projected population growth and future developments are not likely to affect the ability of the disposal facility to meet the performance objectives.
3. Areas shall be avoided having known natural resources which, if exploited, would result in failure to meet the performance objectives.
4. The disposal site shall be generally well drained and free from areas of flooding or frequent ponding. Waste disposal shall not take place in a 100-year floodplain, coastal high-hazard area or wetland, as defined in Executive Order 11988, "Floodplain Management Guidelines."
5. Upstream drainage areas shall be minimized to decrease the amount of runoff which could erode or inundate disposal units.
6. The disposal site shall provide sufficient depth to the water table that groundwater intrusion, perennial or otherwise, into the waste will not occur.
7. Areas shall be avoided that are the recharge areas of sole source aquifers unless it can be demonstrated with reasonable assurance that the disposal site will be designed, constructed, operated, and closed without an unreasonable risk to an aquifer.
8. The hydrogeologic unit used for disposal shall not discharge groundwater to the surface within the disposal site.
9. Areas shall be avoided where tectonic processes such as faulting, folding, seismic activity, or vulcanism may occur with such frequency and extent to significantly affect the ability of the disposal site to meet the performance objectives or may preclude defensible modeling and prediction of long-term impacts.
10. Areas shall be avoided where surface geologic processes such as mass wasting, erosion, slumping, landsliding, or weathering occur with such frequency and extent to significantly affect the ability of the disposal site to meet the performance objectives or preclude defensible modeling and prediction of long-term impacts.
11. The disposal site must not be located where nearby facilities or activities could adversely impact the ability of the site to meet the performance objectives or significantly mask the environmental monitoring program.
12. The disposal site shall not be located where soil conditions are such that spill cleanup would be impracticable.

Source: TDOH, Part 45, 1983.

Table 29. GUIDELINES FOR SITE SELECTION

Exclusions

1.   Disposal sites shall not be located in the 100-year floodplain, coastal high hazard zone, or wetlands.
2.   The site should be located so that upstream drainage is minimal and is easily managed. This generally indicates an area with an existing grade of five percent or less.
3.   Sufficient depth to the water table should be present so that groundwater intrusion, perennial or otherwise, into the waste will not occur. It is the desire of the Authority to locate the disposal site in an area where at no time will the water table rise to within 50 feet of the trench bottom, assuming that the site will be a conventional shallow land burial operation.
4.   Any groundwater discharge to the surface within the disposal site shall not originate within the hydrogeologic unit used for disposal. It is desirable that the site be located where naturally occurring groundwater discharge is not present.

Inclusions

1.   The proposed site should be readily accessible to state highways, and easy access to rail and/or barge transportation is desirable.
2.   The site should be located where calcareous and/or permeable soils occur.
3.   The site should preferably be located on existing State or federally-owned land to minimize site acquisition problems and cost.
4.   It would be desirable to locate the site on land where an Environmental Impact Statement has been previously prepared.
5.   The site should be located in an area of net evaporation and/or evapotranspiration.
6.   The site should be located such that transportation problems from major radioactive waste generators are minimized.
7.   The site should preferably be located in an area where there is little public opposition.
8.   The site should be capable of being characterized, modeled, analyzed, and monitored.

Source: TLLRWDA, 1982.

Alvarado, 1985).[66] Based on the above criteria, eight of the 15 areas were selected as preferred siting areas.[67]

The objective of the next phase was to identify sites for preliminary site characterization. Because of the difficulty in finding properties that would be sold by voluntary sellers with title to both surface and mineral estates, available land was identified and technical screening was completed at the same time.

Fifty-seven potential sites that met geology and transportation guidelines were identified, and ranked according to the following process:

1. Technical experts from the Authority staff and its contractors defined siting issues and subsidiary factors. Issues were separated into two categories: environmental issues and engineering/economic issues (Table 30 and 31).
2. The technical experts rated each site in terms of each factor. The highest rating in terms of each factor was five, for the most desirable condition; and the lowest rate was one, for the least desirable condition.
3. These experts assigned importance weights to factors within each issue. Factor weights were normalized so that they sum to 1.0 for any particular issue (Table 30 and 31).
4. Each issue was assigned an importance weight according to its perceived importance in site design and performance. The scale was one to ten, from least to most important. The importance weights for issues were defined in a two-day, twelve-person workshop attended by the members of the Authority, its contractors, and two representatives of the Citizens' Advisory Panel. Members of the group voted individually. Voting was followed by a discussion of the rationale behind extremely high and low votes. Voting was repeated and discussed until the range of opinion narrowed or the average between successive votes did not change substantially (TLLRWDA, 1990). Weights for environmental and engineering/economic issues were defined separately (Tables 30 and 31).
5. Ratings were combined with the weights for factors and issues to produce overall site ratings. Environmental and engineering/economic ratings were calculated separately.

---

[66] These included geology, groundwater, surface water, soils, mineral and energy resources, meteorology, and historical and archaeological sites.

[67] These were Hudspeth County, Nueces Plains, Black Prairie, North-Central Prairies, Western Cross-Timbers, Abilene-Haskell Plains, Red Bed Plains, and Northern Black Prairie (TLLRWDA, 1990).

Table 30. SUMMARY OF ENVIRONMENTAL ISSUES, FACTORS,
AND FACTOR WEIGHTS FOR RATING OF POTENTIAL SITES*

Issue	Factor	Factor Weight
Historical/Archaeological/ Cultural (4.2)	A-Distance to nearest school/church	0.55
	B-Distance to nearest historical site	0.25
	C-Distance to nearest archaeological site	0.20
Demography (3.7)	A-Dwellings within 1 mile	0.40
	B-Distance to nearest population center	0.25
	C-Distance to nearest town	0.25
	D-Distance to nearest city	0.10
Environment and Recreation (3.3)	A-Distance to wildlife areas	0.30
	B-Distance to state/federal park	0.30
	C-Distance to county parks	0.30
	D-Distance to nearest trail	0.10
Geology (8.4)	A-Complexity	0.55
	B-Stability	0.20
	C-Soil shrink/swell potential	0.15
	D-Distance to unfavorable geology	0.10
Groundwater (9.9)	A-Presence of shallow groundwater	0.75
	B-Presence of underlying aquifer	0.10
	C-Distance to nearest well/windmill	0.10
	D-Groundwater sources within 2 miles	0.05
Hydrometeorology (7.4)	A-Evapotranspiration	0.65
	B-Frequency of severe storms	0.35
Land Use Potential (5.0)	A-Land use on site	0.80
	B-Mineral resource potential	0.20
Nonconforming Land Use (3.5)	A-Proximity	1.00
Surface Water (6.6)	A-Upstream drainage	0.35
	B-Linear stream length within 1 mile	0.20
	C-Downstream tanks/ponds	0.20
	D-Distance to nearest reservoir	0.15
	E-Distance to nearest river	0.10
Transportation (3.8)	A-Total population passed on limited access highway	0.50
	B-Distance from limited access highway	0.30
	C-Distance to LLW generation center	0.20

*Note: The importance weight for each issue is shown in parentheses.
Source: TLLRWDA, 1990.

Table 31. SUMMARY OF ENGINEERING/ECONOMIC ISSUES,
FACTORS, AND FACTOR WEIGHTS FOR RATING OF POTENTIAL
SITES*

Issue	Factor	Factor Weight
Support and Services (5.2)	A-Distance to services	1.00
Land Availability (9.3)	A-Surface rights	0.50
	B-Mineral rights	0.50
Land Costs (1.3)	A-Property costs	0.40
	B-Number of owners	0.40
	C-Mineral rights	0.20
Site Engineering (4.1)	A-Earthwork/excavation difficulty	0.50
	B-Regrading	0.25
	C-Site clearing	0.25
Surface Water Management (3.0)	A-Precipitation	0.75
	B-Upstream drainage	0.25
Transportation Access Costs (6.9)	A-Road access construction	0.40
	B-Freeway access or bridge cost	0.60
Transportation Operating Costs (2.1)	A-Distance from LLW generation center	1.00

*Note: The importance weight for each issue is shown in parentheses.
Source: TLLRWDA, 1990.

6. Sites were rank-ordered considering both the environmental and the
    engineering/economic ratings. Six sites available for purchase were
    selected for preliminary characterization.[68]

---

[68] These included a site in Northeast McMullen County, a site near Asherton in Dimmit
County; a site southwest of Dell City in Hudspeth County (coded as HA-1A0; a site ten
miles south of HU-1A in Hudspeth County (coded as HU01B0; a site near Encinal in
LaSalle County; and a site south of Kalgary in Garza County.

In early 1984, the Keystone Process (Keystone Center, 1982) was used in Dimmit and Hudspeth counties. Citizen review committees were organized in both counties, consisting of concerned citizens appointed by local officials (TLLRWDA, 1985). The committees were asked to provide timely citizen input; identify important issues; provide a dialogue between the Authority and the public; and develop a fact-finding report.

About 50 people were involved in the Keystone Process, and a report was produced that described citizens' concerns in Hudspeth County. Interviews with local and State government representatives, however, suggest that the Keystone exercise did not produce acceptance of the site. On the contrary, by increasing awareness of the siting process, it reinforced most participants' resolve to fight the site.

Site-specific data were collected about the geology, groundwater and surface hydrology, meteorology, biology, and archeology of the six potential sites. In accordance with the statute, the Authority also investigated socioeconomic impacts of the potential sites.[69]

Throughout the site-selection process, the Authority lacked the power of eminent domain. Many sites were rejected because of problems finding landowners who owned both the surface and the mineral estates and were willing to sell them. The Texas General Land Office denied the Authority access to investigate State-owned lands. All State-owned lands were, therefore, excluded from consideration.

In August 1984, the Authority selected two preferred sites located in McMullen County and Dimmit County, in south Texas.

By January 1985, the Authority was planning to name the McMullen site as the most suitable site for LLRW disposal. News of these plans led to a vigorous protest in McMullen County, particularly by citizens of Corpus Christi, a city located about 100 miles from the proposed site. Opponents pointed out the proximity of the site to the Choke Canyon Reservoir, which provides drinking water for the City (Mathews, 1985). The Legislature passed a law (Texas, State of, Ch. 692, 1985) that prohibited siting within 20 miles upstream from a reservoir, thereby making the McMullen County site unsuitable. The amendments also directed the Authority to give preference to State-owned land.

In February 1985, the Authority began evaluating all State-owned land. With the cooperation of the General Land Office and the University of Texas System, the Authority identified the fee-simple lands administered by the

---

[69] Studies of socioeconomic factors conducted for the Authority included: an economic impact analysis by the Bureau of Business Research, University of Texas at Austin (TLLRWDA, 1984/a); a socioeconomic profile of potential areas by the Department of Rural Sociology, Texas A & M University (TLLRWDA, 1984/b); a public opinion poll by the Public Policy Resources Laboratory, Texas A & M University (TLLRWDA, 1984/d); and an analysis of the impact of the site on local governments by the Texas Advisory Commission on Intergovernmental Relations (TLLRWDA, 1984/c).

State of Texas. Next, a process similar to the initial screening (see the section on Statewide Screening) was applied, and three preferred siting areas were identified.[70]

In the next stage, a regional screening of the preferred siting areas identified potential sites using criteria that included geology; groundwater; surface water; accessibility; nonconforming land use; and demographic, cultural, and recreational factors (TLLRWDA, 1990).

Using the results of the regional screening, 107 sites in six counties were identified. In order to evaluate the sites, air and ground reconnaissance studies were performed in June 1985. The reconnaissance studies provided further information about surface soil types, topography, local geology, plants and animals, the location of residences, water wells, water tanks and ponds, site accessibility, current land use, and drainage.

Based on these studies, 18 potential sites were selected by a joint evaluation team consisting of representatives of the Authority, Dames & Moore, the University of Texas Lands Division, and the General Land Office Uplands Division. Two additional sites in Hudspeth County (HU-1A and HU-1B) that had been extensively characterized in the previous study were also incorporated in the ranking process. In a site-ranking process similar to that previously described, seven sites were identified for preliminary site evaluation (TLLRWDA, 1990).[71]

Site-specific investigations were performed at each of the seven sites from July 1985 to November 1986. In November 1986, the Authority named the Fort Hancock and one of the Dell City sites as preferred sites. The Authority prepared an evaluation of costs, socioeconomic impacts, environmental impacts and public health impacts for these two sites (TLLRWDA, 1986/a). In February 1987, the Fort Hancock site was designated for detailed characterization.

The site was located in southern Hudspeth County, approximately 40 miles east of the city of El Paso and ten miles northeast of Fort Hancock. The population of Hudspeth County is about 2,000, with a population density of about one per 1,000 square miles. The County government supported the facility, primarily because of the impact funds established in the statute. Although some local citizen groups in the Fort Hancock area opposed the facility, they never became a vocal majority. El Paso County, however, vigorously opposed the facility.

---

[70] These were: Northern Trans-Pecos, containing Culberson and Hudspeth counties; Southern Trans-Pecos, containing Brewster and Presidio counties; and Southern High Plains, containing Andrews, Loving, Ward, and Winkler counties (TLLRWDA, 1990).

[71] These were: a site near Fort Hancock in southern Hudspeth County; two sites southwest of Dell City in central Hudspeth County; a site in Eastern Culberson County; a site in central-western Culberson County; a site in Loving County; and a site in Ward County.

After designating the Fort Hancock site, the Authority proposed to establish a West Texas Policy Makers Forum, coordinated by the West Texas Council of Governments, that would involve all key stakeholders in a joint evaluation of the proposed site. After initially agreeing, El Paso County vetoed the proposal (Fowler, 1989).

In February 1987, El Paso County filed a lawsuit against the Authority; and, in March 1987, a state district court granted a temporary injunction halting site selection. After the law had been amended in 1987, the 8th Court of Appeals reversed the temporary injunction, and site characterization proceeded. Hydrology, geology, meteorology, topography, archaeology, and other aspects of the proposed Fort Hancock site were addressed in detail (TLLRWDA, 1990).

In 1987, the West Texas Council of Governments (later the Rio Grande Council of Governments) set up technical advisory committees and initiated a geotechnical investigation. The investigation, funded by El Paso County, found the Fort Hancock site technologically unsuitable due to the potential for earthquakes and the danger of water contamination.

After the Authority announced the Fort Hancock site as the preferred site in November 1989, El Paso County reinstated the lawsuit. In January 1991, the State district court in Hudspeth County found the Fort Hancock site unsuitable and issued a permanent injunction.

After the court's decision, negotiations started involving the representatives of the Authority and Hudspeth and El Paso counties. A compromise was reached that a site in eastern Hudspeth County, at least 100 miles from El Paso, would be acceptable for each party. A new bill was drafted that defined a geographic area of 400 square miles in eastern Hudspeth County, where the site must be located. The bill gave the Authority the right of eminent domain. The law was passed in May 1991 (Texas, State of, Ch. 402, 1991).

In August 1991, the Authority selected a disposal site in the designated area. The site is located on a large block of private land (16,000 acres) called Fasken Ranch, seven miles southeast of Sierra Blanca, in the Chihuahua Desert.

In February 1992, the Board of Directors granted approval to file an application for a license to operate a disposal facility. The license application was submitted by the Authority to the Texas Water Commission (TLLRWDA, 1992).

In April 1992, a public hearing was conducted in Sierra Blanca. In May 1992, the Authority's Board of Directors formally designated the Fasken Ranch site, and in June, the Authority purchased the 16,000-acre property. Detailed site characterizations started in July 1992 (TLLRWDA, 1992).

PERMANENT DISPOSAL METHOD SELECTION

The 1981 law did not address the question of disposal method. At that time, it was generally assumed that shallow land burial would be used for waste disposal.

In December 1982, the Authority entered into a contract with Ebasco Services, Inc., to prepare a conceptual design for a disposal facility. Results of this study were published in December 1983. The design indicated that a 200-acre facility could serve Texas for 30 years. The study recommended improved shallow land burial, where each disposal unit incorporates a three-foot-thick clay cup, a two-foot-thick liner, and a total cover of 16 feet over the buried wastes (TLLRWDA, 1983/a).

During this time, however, the public raised questions about: the safety of shallow land burial; proposed engineered, monitorable structures; and retrievable disposal. Many groups favored above-ground disposal.

In 1985, the Legislature amended the statute to ban both traditional and improved forms of shallow land burial. The law mandated that the waste must be contained in a reinforced concrete barrier or in containers constructed of material equivalent or superior to reinforced concrete.

Some legislative leaders pressured for above-ground storage but were defeated. The new language instructed the Authority to consider "alternative management techniques, including aboveground isolation facilities" (Texas, State of, Ch. 692, Section 3.05, 1985). The amendments directed that no waste may be disposed of in a landfill below the natural level of the site unless State or federal programs preclude or recommend against above-ground disposal, or the Authority determines that below-ground disposal provides greater protection than above-ground disposal.

In 1985, the Authority launched a study to investigate alternative disposal technologies conducted jointly by the staff of the Authority and Rogers and Associates Engineering Corporation (TLLRWDA, 1987).

In the first stage, various disposal technologies were reviewed.[72] For comparative purposes, two additional disposal methods excluded by law, shallow-land burial and improved shallow land burial, were also analyzed. The technologies were compared and rank-ordered by using a multiple-criteria evaluation model (TLLRWDA, 1987).

---

[72] The Authority selected the following technologies for further investigation: above-ground vaults; below-ground vaults; above-ground modular concrete canisters; above-/below-ground modular concrete canisters; below-ground modular concrete canisters; earth-mounded concrete bunkers; mined cavities; unlined augered holes; and lined augered holes (TLLRWDA, 1987).

1. Issues and factors of evaluation were determined by the contractors of the Authority, based on Texas (TRCR, Parts 45 and 21) and federal (10 CFR 61 and 10 CFR 20) regulations (Table 32).

2. A two-day workshop was organized in Hudspeth County where a group of local citizens discussed the issues and factors and assigned importance weights to them. Weights of issues ranged from one (least important) to ten (most important); the sum of factor weights for a given issue was 1.0 (Table 33).

3. Technical experts independently rated the performance of each candidate disposal technology according to each factor. Each disposal technology was assigned a rating ranging from one (least attractive) to five (most attractive).

4. Overall ratings were calculated for each disposal technology by considering the factor ratings and the weights on issues and factors (Table 34).

Based on the results of the above analysis and the advice of the Citizens' Advisory Panel, the Authority identified four disposal technologies for further assessment (TLLRWDA, 1990): below-ground modular concrete canisters; below-ground vaults; above-/below-ground modular concrete canisters; and above-ground vaults. Above-ground vaults were included despite an overall low ranking because the statute (Texas, State of, Ch. 692, 1985) specifically prescribed the evaluation of above-grade disposal methods (TLLRWDA, 1990).

Conceptual designs for the methods were developed and investigated. The analysis concluded that all the conceptual facilities are licensable, although there is less certainty about above-ground vaults than about the other technologies. Several regulators and experts questioned whether the above-ground facility can resist various natural and man-made stresses (TLLRWDA, 1990).

Considering the advisory opinion of the Bureau of Radiation Control, the Authority selected below-ground modular concrete canisters as the disposal technology (TLLRWDA, 1990).

COMPENSATION AND INCENTIVES

The Texas LLRW Disposal Authority Act (Texas, State of, Ch. 273, 1981) authorized the Board of Directors to reimburse a city, county, or any other political entity for actual or anticipated expenses occurring as a result of the disposal facility.

Table 32. ISSUES AND FACTORS USED IN DECISION ANALYSIS OF LOW-LEVEL RADIOACTIVE WASTE DISPOSAL TECHNOLOGIES

	Issue		Factor
1.	Radiation Safety to the General Public	1.1	Degree of Protection to Collective Populations
		1.2	Degree of Protection to Critical Population Groups
		1.3	Period of Total Containment
		1.4	Uncertainty of Long-Term Safety Performance Assessment
2.	Worker Safety	2.1	Worker Radiation Doses
		2.2	Industrial Safety
3.	Radiation Safety to Intruders	3.1	Protection of Inadvertent Intruders
		3.2	Attractiveness to Inadvertent Intruders
		3.3	Impacts to Scavengers
4.	Economics and Cost	4.1	Unit Disposal Cost ($ per cubic foot)
		4.2	Economic Benefit to Local Community
		4.3	Economic Cost to Local Community
5.	Remedial Action	5.1	Cost of Remedial Action
		5.2	Radiation Exposures During Remedial Action
		5.3	Expected Effectiveness of Remedial Action
6.	Ease of Implementation	6.1	Regulatory Tools Available for Licensing
		6.2	Uncertainty in Meeting Licensing Requirements
7.	Operational Flexibility	7.1	Ability to Accept all Waste Forms and Packages
		7.2	Disposal Operations Sequencing
		7.3	Construction Activities Management
8.	Post-Closure Stability	8.1	Period of Active Maintenance
		8.2	Effectiveness Against Natural Disruptions
9.	Impact on Local Community	9.1	Aesthetics of Facility
		9.2	Number of People Working at Facility
		9.3	Long-Term Cultural and Social Impacts
10.	Site and Facility Considerations	10.1	Cost Effectiveness of Siting Facility
		10.2	Feasibility of Siting Facility
11.	Non-Radiological Environmental Impacts	11.1	Emissions from Facility
		11.2	Impacts from Related Activities
		11.3	Ease of Effluent Control

Source: TLLRWDA, 1987.

Table 33. ISSUE AND FACTOR IMPORTANCE WEIGHTS USED IN THE DECISION ANALYSIS

Issue/Factor	Weight	
1. Radiation Safety to the General Public	10.0	
1.1 Degree of Protection to Collective Populations		0.3
1.2 Degree of Protection to Critical Population Groups		0.4
1.3 Period of Total Containment		0.2
1.4 Uncertainty of Long-Term Safety Performance Assessments		0,1
2. Worker Safety	10.0	
2.1 Worker Radiation Doses		0.5
2.2 Industrial Safety		0.5
3. Radiation Safety to Intruders	6.6	
3.1 Protection of Inadvertent Intruders		0.2
3.2 Attractiveness to Inadvertent Intruders		0.2
3.3 Impacts to Scavengers		0.3
3.4 Security Against Scavengers		0.3
4. Economics and Cost	9.8	
4.1 Unit Disposal Cost ($ per cubic ft)		0.3
4.2 Economic Benefit to Local Community		0.4
4.3 Economic Cost to Local Community		0.3
5. Remedial Action	6.4	
5.1 Cost of Remedial Action		0.3
5.2 Radiation Exposures During Remedial Action		0.3
5.3 Expected Effectiveness of Remedial Action		0.4
6. Ease of Implementation	7.8	
6.1 Regulatory Tools Available for Licensing		0.6
6.2 Uncertainty in Meeting Licensing Requirements		0.4
7. Operational Flexibility	7.9	
7.1 Ability to Accept all Waste Forms and Packages		0.4
7.2 Disposal Operations Sequencing		0.2
7.3 Construction Activities Management		0.4
8. Post-Closure Stability	7.2	
8.1 Period of Active Maintenance		0.6
8.2 Effectiveness Against Natural Disruptions		0.4
9. Impact on Local Community	9.7	
9.1 Aesthetics of Facility		0.2
9.2 Number of People Working at Facility		0.3
9.3 Long-Term Cultural and Social Impacts		0.5
10. Site and Facility Considerations	9.2	
10.1 Cost Effectiveness of Siting Facility		0.5
10.2 Feasibility of Siting Facility		0.5
11. Non-Radiological Environmental Impacts	7.9	
11.1 Emissions from Facility		0.3
11.2 Impacts from Related Activities		0.5
11.3 Ease of Effluent Control		0.2

Source: TLLRWDA, 1990.

Table 34. ISSUE AND TOTAL WEIGHTED SCORES FOR ELEVEN DISPOSAL TECHNOLOGIES

	Improved Shallow Land Disposal	Shallow Land Disposal	Below-Ground Vault Disposal	Below-Ground Modular Concrete Canister Disposal	Lined Augered Holes	Unlined Augered Holes
Radiation Safety to the General Public	28	23	35	36	33	29
Worker Safety	40	40	33	28	33	33
Radiation Safety to Intruders	18	15	22	17	24	24
Economics and Costs	35	35	30	29	29	29
Remedial Action	18	18	22	18	16	15
Ease of Implementation	37	38	26	19	25	25
Operational Flexibility	35	35	23	18	19	21
Post-Closure Stability	22	19	22	18	25	25
Impact on Local Community	35	34	33	25	31	30
Site and Facility Considerations	31	32	29	29	25	25
Non-Radiological Environmental Impacts	29	29	23	31	26	25
Totals	329	319	229	297	286	281

Table 34 Continued

	Above-/Below-Ground Modular Concrete Canister Disposal	Above-Ground Modular Concrete Canister Disposal	Earth-Mounded Concrete Bunker Disposal	Mined Cavity Disposal	Above-Ground Vault Disposal
Radiation Safety to the General Public	31	26	32	36	21
Worker Safety	27	26	25	19	27
Radiation Safety to Intruders	19	18	17	26	14
Economics and Costs	27	27	29	25	29
Remedial Action	22	23	18	11	24
Ease of Implementation	17	16	19	13	12
Operational Flexibility	21	22	18	21	22
Post-Closure Stability	21	24	18	29	10
Impact on Local Community	25	23	25	37	25
Site and Facility Considerations	29	30	29	13	29
Non-Radiological Environmental Impacts	21	20	21	25	22
Totals	260	253	251	245	234

Source: TLLRWDA, 1990.

In 1987, the Legislature passed an impact assistance bill (Texas, State of, Ch. 76, 1987). It provided that, to the extent possible, all purchases of services and equipment be made locally and all workers used during construction and operation be hired locally. The law also directed that at least ten percent of the gross revenues of the site be provided to compensate the affected communities, approximately $1,000,000 to $1,300,000 annually. According to the law, the local Citizens' Advisory Committee will make recommendations to the Board of Directors concerning requests for assistance. The request has to be approved by the Board. State and local government officials, as well as local citizens, criticized this compensation as highly inflexible and excluding compensation for socioeconomic damages.

In 1991, amendments to the law were passed, allowing Texas to enter an interstate compact (Texas, State of, Ch. 583, 1991). In the event of such a compact, the amendments provided additional compensation for affected communities. Under the new law, the host county will receive ten percent of the money paid by another state wishing to compact with Texas or $1 million (whichever is more).

INTERIM WASTE MANAGEMENT

The LLRW Disposal Authority Act of 1981 included no provisions for interim waste management. By 1983, however, it became important to investigate interim storage in case the state couldn't provide for permanent LLRW disposal by January 1, 1986. In 1983, the Authority entered into a contract with Ebasco Services, Inc., to perform a study on the design of an interim storage facility. The facility design was based on a concept developed for the interim storage of waste at nuclear power plants. The facility design incorporated a reinforced concrete structure on a reinforced concrete basement (TLLRWDA, 1983/b).

The amendments of 1985 directed the Authority to investigate waste storage, and the 1987 amendments required the nuclear power industry to provide storage facilities at reactor sites sufficient for five years (Texas, State of, Ch. 692, 1985; Ch. 22, 1987).

In 1989, the Authority surveyed major LLRW generators concerning their plans for waste disposal after 1992, or until a Texas disposal facility was expected to be available. The survey showed that many of the industrial and institutional generators planned to use on-site storage. Some indicated that they had to establish additional storage facilities. Others planned to get their waste stored, treated, and processed by brokers (Clements, 1989).

By June 1991, all generators were required to develop and submit to the Department of Health detailed plans for storing or otherwise managing their waste until the Texas disposal facility is operational. The Department of

Health planned to issue any needed licenses and license amendments by December 31, 1992.

ANALYZING THE TEXAS CASE HISTORY FROM THE PERSPECTIVE OF THE COMPETING VALUES APPROACH

Texas was among the quickest states to respond to the 1980 federal LLRW Management Act. By 1981, the State had decided not to join a compact but to "go it alone" and to establish a LLRW disposal facility for its in-state generators. From the rational perspective, for the next several years, the Texas siting process appeared to be perhaps the strongest in the nation. The process was sharply goal-focused and serious efforts to find an appropriate site for the facility were clearly underway.

As it became increasingly clear that only a few states were making genuine progress toward siting a new facility, however, concern developed that if Texas opened a site before other states did so, it could become de facto a national facility. In 1985, therefore, the legislature enacted a law directing the State to deliberately delay the siting. Since then, siting efforts progressed at a considerably slower pace. In 1991, however, a site was selected near Sierra Blanca and, in 1992, a license application was submitted.

The Texas process was strong from the empirical, as well as the rational, perspective. Texas placed considerable emphasis on conducting a data-based siting process and achieving accountability in the final decision. An independent agency, the LLRW Disposal Authority was established to select a site and a disposal method, construct and operate the facility, develop a compensation and incentive package, and provide for interim management. Decisions of the Authority were extensively documented. To help to strengthen the scientific integrity of the studies and the accountability of the decisions, the Authority involved a large number of private contractors, State agencies, and organizations in the Texas university system.

Large amounts of data about geology, hydrology, meteorology, archeology, and other technical characteristics of the investigated sites were collected. State-sponsored studies also investigated socioeconomic factors, including impacts on local governments, economic impacts, and public opinion near candidate sites. Significant amounts of technical information were collected about alternative disposal technologies, as well as storage facility design.

The Texas LLRW Disposal Authority Act dictated the Authority to select the "most suitable site." To identify optimal sites and methods, the Authority used complex, hierarchical MAU models. Some members of the public challenged the site selection criteria, however, and, in response to vigorous public opposition near candidate sites, in 1985, the Texas legislature instructed the Authority to modify its site selection criteria. Subsequently, the new site selected by the Authority was legally challenged by local municipalities for not being the most suitable site. In 1987, Texas legislature amended the law to drop the provision that prescribed selection of the technically optimal site.

With regard to the consensual perspective, which emphasizes participatory processes and supportable decisions, Texas law instructed the Authority to invite local government officials and members of other local organizations to participate in the siting process. In compliance, the Authority made several efforts to encourage public participation.

In 1983, a six-member Citizens Advisory Panel was created. The Panel included citizens from all regions of the state, and its purpose was to help to identify site and method selection criteria and assign relative weights to them. Members of the Panel, however, participated only in a few steps of the procedure. The Panel did not appear to contribute significantly toward the development of public support for the process.

In 1984, the Authority tried the Keystone Process, a conflict resolution method based on soliciting early public input, in two counties where candidate areas were located. In 1987, the Authority proposed establishing a forum that would involve affected local governments and other key stakeholders in a joint evaluation of the selected Fort Hancock site. These attempts failed to give local communities any power to influence the location of the site and were therefore greeted with skepticism. In 1991, however, local government officials played instrumental roles in discussions that resulted finally in a legislative agreement on a siting area acceptable to all the key parties.

Finally, with respect to the political perspective, the Texas process placed little emphasis on adaptability. The LLRW Disposal Authority Act provided extensive, detailed instructions governing the site- and method-selection processes, including the requirement that only the optimal site and method be selected. The Authority developed detailed plans and tried to follow them, but the candidate sites selected by the Authority were greeted with strong public and political opposition.

Although the initial legislation provided relatively little room for flexibility, the process ultimately proved quite adaptable and eventually yielded a substantial level of political legitimacy. This was accomplished primarily by repeatedly amending the law in response to public opposition and legal challenges to the siting process. The amendments of 1985 directed the Authority to focus on state-owned land and expressed preference for above grade disposal. The amendments of 1987 specified a compensation scheme. In 1991, legislators and local governments negotiated an agreement

on a siting area as well as an incentive package, which was then enacted into
State law.

Chapter 8

CANADA

BACKGROUND

In Canada, LLRW is defined as all radioactive wastes except for irradiated nuclear fuel from nuclear reactors and by-products of uranium mining and milling. LLRW also includes so-called "historic wastes," originating from earlier activities, primarily involving the radium industry, that are managed in a manner no longer considered acceptable and for which the original producer cannot be reasonably held responsible (Morrison & Brown, 1991).

In Canada, generators bear primary responsibility for the management of LLRW. LLRW produced by utilities are stored in storage facilities maintained by nuclear power plants. Non-utility LLRW is shipped to a storage facility of Atomic Energy Canada Limited, a national research institution, at the Chalk River Nuclear Laboratories.

Historic waste constitutes about 90 percent of the current LLRW inventory, estimated at 1,000,000 cubic meters (Morrison & Brown, 1991). Siting efforts have focused on finding a suitable site and method for disposing of historic wastes accumulated primarily in the province of Ontario.[73]

From 1932 to 1988, a radium and, later, uranium refinery was operated in the Port Hope area on Lake Ontario by the federal government as a Crown Corporation called Eldorado Resources Ltd. (Eldorado). Initially, radioactive wastes were disposed of on the plant site; later, in municipal landfill sites, in ravines, and in the harbor of Port Hope. From 1948 to 1955, the wastes were disposed of in a LLRW disposal facility developed by Eldorado at the Welcome site near Port Hope. From 1955 the wastes were disposed of at a second Eldorado site located at Port Granby, on the bluffs above Lake Ontario (Morrison & Brown, 1991).

In 1975, it was discovered that buildings, soil, and groundwater in the Port Hope area had become widely contaminated due to haphazard disposal. At the Port Granby site, the bluffs have eroded, threatening the continued containment of wastes (Morrison & Brown, 1991).

---

[73] The estimated amount of historic waste accumulated in the Port Hope area is 880,000 cubic meters (Armour, 1991).

A chronological summary of key events related to the LLRW disposal facility siting in Canada is given in Table 35.

Table 35. CHRONOLOGICAL SUMMARY OF KEY EVENTS IN CANADA

Date	Event
1980	The Atomic Energy Control Board directed Eldorado Resources Ltd. (Eldorado) to decommission two LLRW disposal facilities.
1984	Eldorado identified Port Granby and Lake Shore Road (Newcastle) and Wesleyville (Hope Township) as potential sites.
1985	The federal government ordered Eldorado to identify sites at least five kilometers from Lake Ontario.
1986	Eldorado identified Newcastle North, Newcastle East, and Wesleyville as potential sites. The federal government changed its policy with regard to historic waste; ordered Eldorado to stop the ongoing siting process; and established a seven-member Siting Process Task Force to recommend a siting process to be conducted by the government.
1987	The Siting Process Task Force recommended a process based on joint problem solving with volunteer communities.
1988	A Siting Task Force was formed to implement the planning phase of the siting process.
1989	Fourteen communities expressed interest in being considered for hosting the facility.
1990	Four municipal councils passed resolutions stating their interest in participating in the next phases of the process. The Siting Task Force recommended Deep River, Geraldton, and Hornepayne (volunteer communities), and Port Hope, Hope Township, and Newcastle (source communities) as potential sites and long-term storage as preferred waste management method.
1992	A new Siting Task Force was established to assess and implement the project.

In 1976, the federal government and the government of Ontario established a joint task force to investigate contamination in the Port Hope area and to recommend remedies. From 1977 to 1981, about 72,000 cubic meters of contaminated soil and building materials were removed by the government from the Port Hope area and transported to the Chalk River site.

OVERALL APPROACH

In 1980, the Atomic Energy Control Board (AECB), the federal regulatory agency, directed Eldorado to develop plans to decommission the Port Granby site. Later, the AECB instructed the corporation to decommission the

Welcome site because of the possibility of future groundwater contamination. In 1982, Eldorado announced that it would decommission both sites, establish a new disposal facility and remove the waste from the Welcome and Port Granby sites, together with some waste and contaminated soil from Port Hope, to the new site.

From 1982 to 1986 Eldorado attempted to find a site for the disposal facility in the Port Hope area. Because of vigorous opposition to the siting by nearby communities, in 1986 the federal government modified its policy on LLRW management.

In April 1986, the Minister of Energy, Mines and Resources released a statement outlining the new federal policy (Canada Department of Energy, Mines and Resources, 1986/a).

The statement confirmed that the primary responsibility for LLRW management should rest with waste generators. It concluded, however, that the federal government should accept residual responsibility when no person or company can be held responsible for the wastes.[74]

In December 1986, because of the public protest, the federal government ordered Eldorado to stop the siting process. The Minister of Energy, Mines and Resources authorized establishment of a seven-member Siting Process Task Force, an independent government body, to recommend a process for siting a disposal facility for existing, on-going, and historic wastes in the Port Hope area (Canada Department of Energy, Mines and Resources, 1986/a). The Minister appointed the members of the Task Force, including a government officer, a resident from the Port Hope area, and five experts in nuclear medicine, industrial engineering, urban and regional planning, environmental management, and sociology.

The Siting Process Task Force prepared recommendations for a voluntary and cooperative siting process. Based on its recommendations, in 1988, a five-member Siting Task Force was formed to implement the planning phase of the siting process. Its members included a local government official, a resident from the affected region, and three experts in nuclear medicine, urban and regional planning, and sociology. After completion of the planning phase, a new Siting Task Force was established in 1992. As of December 1992, the siting process was in progress.

---

[74] Government responsibility was deemed appropriate for: the cleanup and disposal of historic wastes; wastes generated by small producers who are not in a position to develop disposal facilities; wastes generated by companies no longer in business; and long-term stewardship of disposal sites after they have been closed to the satisfaction of regulatory authorities (Canada Department of Energy, Mines and Resources, 1986/a).

INSTITUTIONAL STRUCTURE

In Canada, the Department of Energy, Mines and Resources has responsibility for managing nuclear material, whereas the AECB has responsibility for "regulating all aspects of the nuclear industry," including LLRW management facilities.

In 1982, the Minister of Energy, Mines and Resources established the Low-Level Radioactive Waste Management Office administered by Atomic Energy Canada Limited. It is responsible for managing LLRW, including cleanup of contaminated sites.

According to the federal policy, Eldorado originally had primary responsibility for decommissioning its old sites and providing for permanent disposal of the waste. In 1986, responsibility for siting a facility for the disposal of historic waste was taken away from Eldorado, and given to the Siting Process Task Force, and later to the Siting Task Force. Both the Siting Process Task Force and the Siting Task Force received advice on regulatory and technical issues from several federal and provincial agencies, including the AECB, the Low-Level Radioactive Waste Management Office, Environment Canada, Ontario Ministry of the Environment, Canadian Transport Commission, Department of Transport, and Ontario Ministry of Transportation and Communications.

SITE SELECTION FOR THE DISPOSAL FACILITY

In 1982, Eldorado initiated consultations with municipal and county governments in the Port Hope area. The rationale for choosing this area as a potential site was that it possessed suitable geology; transportation would be safer and less expensive because of relatively short distances involved; and public acceptance would be greater because of the communities' familiarity with the problem and their economic dependence on the nuclear industry. In addition, government policy encouraged focusing site-selection efforts on a region within 50 kilometers of Port Hope (Kemp, 1989).

Eldorado formed consultative committees in Newcastle and Hope townships, including members of the municipal and county councils and representatives of the local communities. Committees existed for about a year, but because of strong local opposition to the siting, both committees ended discussions.

In 1984, Eldorado identified three potential sites: the Port Granby and Lake Shore Road sites in the town of Newcastle, and the Wesleyville site in Hope Township. This was followed, however, by vigorous public protest in these communities. Legislators also became involved. In August 1984, before the federal elections, the leader of the opposition promised to fight against siting in either area.

After the elections, the federal government issued a directive, ordering Eldorado to identify sites at least five kilometers from Lake Ontario and north of Highway 401. As a consequence, Eldorado had to drop both the Port Granby and the Lake Shore Road sites and identify two new sites.

In August 1986, Eldorado identified three potential sites: Newcastle North, Newcastle East, and Wesleyville. Later, it dropped the Newcastle East site for technical reasons. There was again strong opposition in the affected communities. Petitions against the siting were sent to the federal government, and protest demonstrations were organized (Kemp, 1989).

In October 1986, the federal government directed Eldorado to stop work on its site proposals. The government directive argued that the Newcastle North site was located on valuable agricultural land. The document did not reject the Wesleyville site (Canada Department of Energy, Mines and Resources, 1986/b).

In December 1986, the government halted the site selection and created the Siting Process Task Force to recommend a less confrontational process (Morrison & Brown, 1991).

The Siting Process Task Force began work in January 1987 with an analysis of the technical issues of LLRW management. After an extensive literature review, the Task Force organized a series of round-table discussions with representatives of organizations involved in other siting processes, including the Ontario Waste Management Corporation, the Alberta Special Waste Management Corporation, the Manitoba Department of the Environment, and Ontario Hydro (Siting Process Task Force, 1987).

Next, to gain an insight into public opinion, the Task Force started a broad consultation exercise. Representatives of national and provincial interest groups, and experts from government, industry, and academia were invited to share their experiences with facility siting processes. Communities were visited by the Task Force and asked about the key elements of a siting process acceptable to them (Siting Process Task Force, 1987).[75]

---

[75] The elements regarded as most important and mentioned most frequently by those surveyed included: the community should have power to decide whether to accept a proposed facility; the community should have a direct and active role in decision making, including site monitoring and health monitoring; the "proponent" should be removed from the siting process and an independent group given responsibility to implement the process; the process should be collaborative -- whoever implements the process should have a non-authoritarian/non-paternalistic attitude; to build trust, all stakeholders should be involved in gathering and assessing information; the community should be provided with resources so that it can participate effectively; compensation should be provided to those who are adversely affected, and the community should receive a net benefit; limits on compensation and benefits should be explicit from the outset; and costs should be borne by the waste generators (Armour, 1991).

Based on the input generated by the public, in December 1987, the Task Force released the report entitled "Opting for Cooperation" (Siting Process Task Force, 1987). The report recommended a siting process where communities had to volunteer to be considered as a site and could opt out of the process at any time. In addition, the report recommended that the community should have the status of a partner in problem-solving and decision making throughout the process and should receive compensation to offset unmitigable impacts and to enhance local benefits. The community should also have the right to select, from the given technical options and impact management measures, those acceptable to it. Finally, the report recommended that the Siting Task Force should work with the responsible regulatory agencies to ensure that the safety of the environment and human health were not compromised for any reason. In order to avoid exploiting economically and socially disadvantaged communities, the Task Force also recommended adopting several safeguards.[76]

The Task Force recommended a five-phase siting process (Siting Process Task Force, 1987):

> Phase One: Establishing Guidelines;
> Phase Two: Conducting Regional Information Sessions;
> Phase Three: Informing and Consulting with the Community;
> Phase Four: Assessing the Project; and
> Phase Five: Implementing the Project.

Acting on the recommendations of the Siting Process Task Force Report, in September 1988, the Minister of Energy, Mines and Resources appointed the new Siting Task Force to implement the first three phases of the siting process and to report its findings regarding potential volunteer communities in 18 months (Siting Task Force, 1990/c).

Initiating Phase One, the Siting Task Force reviewed available information on LLRW accumulations and current management practices in Ontario. The Task Force determined that its effort should focus on the historic LLRW accumulations located in Port Hope and Scarborough, and at the Welcome and Port Granby sites in Hope Township and Newcastle (Siting Task Force, 1990/c).

---

[76] These safeguards included: using an explicit, up-front impact management policy to ensure that all communities are aware of the range of options available to them; hiring advisors selected by the community to ensure that local interests are protected throughout the cooperative process of fact-finding and problem-solving; funding thorough site and technology assessments so that decisions by the community, the Siting Task Force, and technical advisors will be based on full information; establishing a broadly based Community Liaison Group to work with the Siting Task Force and local community officials; and, providing funding to allow communities to participate in the process (Armour, 1991).

The Task Force drafted guidelines for eliminating sites from further consideration, impact management and compensation, and the establishment of Community Liaison Groups. The Task Force then held a series of public meetings to present the draft guidelines and elicit comments that helped the Task Force to finalize the guidelines (Siting Task Force, 1990/b).

The Task Force identified two levels of criteria for site elimination. Level one criteria are exclusionary and will be applied by the community and the Siting Task Force in conjunction with regulatory agencies to eliminate obviously unsuitable areas. Level one criteria are summarized in Table 36 (Siting Task Force, 1990/a).

Table 36. SITE ELIMINATION CRITERIA: LEVEL ONE CRITERIA

1. Aggregate resources that could be identified by Ontario's Ministry of Natural Resources and the municipality as requiring protection, pursuant to the provincial Mineral Aggregate Resources Policy Statement.
2. Federally or provincially designated Parks, existing and proposed.
3. Natural areas significant to the Province, such as: lands containing flora or fauna which are rare or unique to the Province; critical wildlife habitats; significant wetlands; Areas of Natural and Scientific Interest (ANSI's); Carolinian Canada sites; or other designated sensitive natural areas in their federal, provincial, and regional context.
4. Lands subject to significant seismic activity with the potential to adversely affect the performance of a waste management facility.
5. Any lands located within the floodplain limits established by the regulatory flood standard appropriate to the region, as indicated in the Provincial policy statement on Flood Plain Planning under Section 3 of the Planning Act of 1983.
6. Land indicating the potential for significant subsidence, erosion, solutioning, landsliding, and weathering that could adversely affect the performance of a waste management facility.
7. Sites containing federally- or provincially-designated archaeologic, historic, or cultural resources.

Source: Siting Task Force, 1990/a.

Unlike level one criteria, level two criteria, summarized in Table 37, are not absolute constraints. They require judgment by both the community and the Task Force (Siting Task Force, 1990/a).

Table 37. SITE ELIMINATION CRITERIA: LEVEL TWO CRITERIA

1. A water body which serves local or regional water supply or recreational needs.
2. Prime agricultural land or the viability of local agricultural operations.
3. Sensitive population concentrations that may be disrupted by the noise, truck traffic, or other effects associated with the day-to-day operation of the facility.
4. Important natural resources, such as water, timberlands or wildlife resources, or mineral deposits.
5. Regionally- or locally-valued archaelogic, historic, or cultural resources.
6. The reasonable cost of construction or access to a major transportation route.
7. Any other feature or resource that the community considers to be significant.

Source: Siting Task Force, 1990/a.

Initiating Phase Two, in Spring 1989, the Task Force invited 850 Ontario municipalities to send representatives to one of eight information sessions held throughout Ontario. Local government officials were provided information about the LLRW problem and the siting process. At each session, participants were informed that the Task Force could organize similar information sessions in their communities so that local residents could learn about the process (Siting Task Force, 1990/c). As a result of the information sessions, 26 local governments invited the Task Force to conduct similar sessions in their communities.

As part of Phase Three, from April to September 1989, information meetings were held in 21 communities (five of the 26 communities withdrew before the meetings). At the meetings, the Task Force explained the nature of the LLRW problem and outlined the principles, safeguards, and steps of the siting process. It was emphasized that the Task Force would not consider a community unless invited by the local council and that no community would be asked to volunteer to accept the facility before the last phase of the process. Each community was advised that if it wished to be considered in the siting process, its council should pass a resolution stating its interest and requesting the Task Force to establish a Community Liaison Group consisting of local residents (Siting Task Force, 1990/c; Armour, 1991).

After the community information meetings, seven communities withdrew, and 14 communities expressed interest in being further considered for hosting the facility. Some of the four "source" communities where historic wastes were located also expressed an interest in being volunteers, the same communities that, a few years earlier, had strongly opposed the siting by Eldorado (Siting Task Force, 1990/c; Armour, 1991).

In the next stage, Community Liaison Groups (CLGs) were formed in the volunteer communities and in three of the four source communities.[77] Members of the CLGs had to be residents of the community and have a record of community involvement. Membership had to reflect the broad interests of the community. Local council members were excluded because the councils made the final decisions on the recommendations of the CLGs.

Working with the Task Force, the CLGs were responsible for developing and implementing a community consultation program. Working with the Task Force, municipal council, and regulatory agencies, they were to eliminate technically unsuitable sites. Finally, they were to prepare a report to the Task Force and municipal council containing findings and recommendations (Armour, 1991).[78]

By December 1989, 14 CLGs were formed.[79] After several months of community consultation, CLGs reported their recommendations in April 1990. Table 38 summarizes the outcome of the CLG recommendations and the resolutions of the municipal councils. Not all town councils accepted the recommendations of the CLGs (Siting Task Force, 1990/b).

In one case, the CLG recommended opting out because there was no suitable land within the municipal boundaries for the proposed facility. In other cases, the primary reasons for opting out were residents' concerns about possible adverse impacts on their local economy and tourism, long-term health effects, risks associated with transporting the wastes, and the long-term physical integrity of the facility. In addition, some CLGs felt the north was unfairly becoming the dumping ground for the south and mistrusted the federal government's commitment to provide the required funds to implement the project (Siting Task Force, 1990/b).

By August 1990, four municipal councils, including Deep River, Chalk River, Geraldton, and Hornepayne, had passed resolutions stating their interest in participating in the next phase of the process. Chalk River was

---

[77] No CLG was formed in the fourth source community, Scarborough, because the federal government and the government of Ontario had signed an agreement that the small volume of the waste located in this community would be placed in interim storage at a provincially designated site until a final disposal site was available. The other three source communities, Port Hope, Hope Township, and Newcastle needed to establish CLGs because they did not have the option to reject participation in the process.

[78] Source communities did not need to prepare reports to their councils.

[79] These were located in the potential volunteer communities of Manitouwadge, Ear Falls, Red Lake, Atikokan, Mattice-Val Cote, Upsala, Coalition of four municipalities (Deep River, Chalk River, The United Townships of Rolph, Wylie, Buchanan, and McKay, and the Townships of Head, Clara and Maria), Hornepayne, Elliot Lake, Township of James, and Geraldton, and in the source communities of Port Hope, Hope Township, and Newcastle.

Table 38. RESPONSE FROM POTENTIAL VOLUNTEER
COMMUNITIES

	Community	CLG Recommendation To Continue	Council Resolution To Continue
1.	Manitouwadge	No	No
2.	Ear Falls	No	No
3.	Red Lake	No	No
4.	Atikokan	No	No
5.	Mattice-Val Cote	No	No
6.	Upsala	No	No
7.	Coalition of four municipalities*	No	Yes: Chalk River & Deep
8.	Hornepayne**	No	Yes
9.	Elliot Lake	Yes	No
10.	Township of James	Yes	No Response
11.	Geraldton	Yes	Yes

*Deep River, Chalk River, The United Townships of Rolph, Wylie,
Buchanan and McKay and the Townships of Head, Clara and Maria.
**CLG originally recommended against continuing but supported Council's
resolution.

Source communities
All source communities will continue for the remainder of the Process.

Source: Siting Process Task Force, 1987.

later eliminated by the Task Force because it lacked sufficient suitable land
within its boundaries (Armour, 1991).

In August 1990, the Siting Task Force published its recommendations
(Siting Task Force, 1990/b) concerning: Deep River, Geraldton, and
Hornepayne (volunteer communities), and Port Hope, Hope Township, and
Newcastle (source communities).[80]

The Task Force report included other recommendations with regard to
Phases Four and Five. The most important recommendations included
extending the time to complete the process to 36-40 months, requesting
adequate resources from the federal government, and calling for specific
funding for impact management and compensation.

---

[80] Source communities had to stay in the process through Phase Four.

In August 1990, the report of the Siting Task Force was submitted to the federal Cabinet. In August 1991, the Department of Energy, Mines and Resources announced it was willing to continue sponsoring the siting process and awarded $22 million in funding for the activities of Phase Four. The government accepted all recommendations of the Siting Task Force report and appointed the chairperson of the new Task Force. Members of the new Task Force were not appointed, however, until January 1992.

According to the plans, Phase Four will include two stages. The first stage requires communities to agree on the most appropriate site and method. It will end with a commitment from the municipal council to proceed to more detailed investigation.   In   the   second   stage,   detailed   environmental assessment and site characterization will be done.   In addition, impact management measures and specifics of a compensation and incentive package will be identified. The Siting Task Force report recommended that a binding referendum take place in the candidate communities in Phase Four. Phase Four will be concluded by a council resolution accepting or rejecting the proposed waste management solution (Siting Task Force, 1990/b).

According to the plans, Phase Five will involve transforming the outcome of Phase Four into legal agreements acceptable to the community and the federal government. After the Cabinet decision, a Board of Directors will be established to oversee the design, construction, and operation of the facility. The Board of Directors includes representatives of the community, the waste generators, the federal government, and the provincial government.   The operator of the facility will be hired by and will report to the Board of Directors, while the Board of Directors will report to the Department of Energy, Mines and Resources (Siting Process Task Force, 1987; Siting Task Force, 1990/b).

PERMANENT DISPOSAL METHOD SELECTION

According to the AECB regulatory document, "Regulatory Objectives, Requirements and Guidelines for the Disposal of Radioactive Wastes" (AECB, 1987), waste disposal is "a permanent method of management in which there is no intention of retrieval and which, ideally, does not rely for its success on the continued need for institutional controls beyond a reasonable period of time."   This concept of waste disposal is strongly supported by AECB, although it is not universally accepted, even within the nuclear community.   Many favor long-term storage, which involves the retrievability, monitoring, and ongoing institutional control of wastes (Siting Process Task Force, 1987).

From 1982 to 1984, Eldorado investigated intermediate depth cavern and near-surface, engineered burial mound.   Intermediate depth cavern was significantly more expensive than near-surface, engineered burial. AECB indicated to Eldorado that, while the intermediate depth caverns met the agency's regulatory criteria, the near-surface, engineered burial did not because it required continuing care and maintenance for long-term safety.

The engineered mound also required thicker soil, such as is typically found in farmland. Since Eldorado investigated sites that would be appropriate for the methods, it considered some farmland sites. This elicited strong criticism from both the affected communities and the federal government, and the process was halted before a preferred disposal method was selected.

After Eldorado's attempts to site the facility failed, the Siting Process Task Force investigated various waste management methods. The Task Force supported long-term storage, which "would involve placing wastes in a facility in such a way that active management would not be required, but any migration of wastes from the facility could be monitored, and, if necessary, as a remedial measure, material could be retrieved" (Siting Process Task Force, 1987; p. 26). The Siting Task Force supported this approach because it anticipated that there might be a future need to upgrade facilities in response to advances in technology.

Based on the negative record of shallow land burial, the Siting Process Task Force supported engineered rather than naturally occurring barriers. In 1987, Acres International Limited was hired to investigate alternative engineered technologies. The Acres study investigated eight alternatives that were evaluated in terms of 11 criteria.[81]

The results of the evaluation are shown in Table 39. The Siting Process Task Force concluded that, except for uranium tailings and abandoned mines, all investigated methods are suitable for the management of LLRW. It was not the intention of the Task Force to select a method at this stage because, according to one of the basic principles of the process, candidate communities have the right to select the technical options acceptable to them. Methods will be selected in Phase Four in consultation with the candidate communities.

---

[81] The alternatives were: uranium tailings (i.e., disposal on top of uranium mine tailings in existing facilities); above-ground vault; trench; below-ground vault; modular concrete canister; borehole; mined cavern; and abandoned mine. The criteria included: public safety; worker safety; intruder safety; consequence of failure; siting ease; retrievability ease; long-term predictability; need for institutional control; cost; labour force size; and community acceptance (Acres International Ltd., 1987).

Table 39. A QUALITATIVE RANKING OF CONCEPTS FOR DISPOSAL OF ONTARIO'S LOW-LEVEL RADIOACTIVE WASTES

	Uranium Trailings	Above Ground Vault	Trench	Below Ground Vault	Modular Concrete Canister	Borehole	Mined Cavern	Abandoned Mine
Public safety	*	C	C	B	B	B	A	B
Worker safety	C	B	A	B	B	B	B	C-
Intruder safety	*	B	C	B	B	B	A	B
Consequence of failure	A	B	B	B	B	B	A	A
Siting ease	C	A	B	B	B	B-	B	C
Retrievability ease	A	A	B	B	B+	B	C	C
Long-term predictability	C	B	B	B	B	B	B	C
Need for institutional control	*	C	B	B	B	B	A	A
Cost	A	C	B	B-	B-	B-	C	B-
Labour force size	A	C	B+	B	B	B	C	B
Community acceptance	B	B	B	B	B	B	A	A

A = most favorable

C = least favorable

* = see source of discussion

Source: Acres International Ltd., 1987.

COMPENSATION AND INCENTIVES

During the attempts of Eldorado to site a disposal facility from 1982 to 1986, principles of compensation and incentives were not formalized. Compensation and incentives were offered to affected communities, but negotiations broke down because communities were not willing to host the facility.

The new 1986 federal policy for LLRW acknowledged that "equity may also require that an area that hosts a safe disposal facility obtain offscts for doing so" (Canada Department of Energy, Mines and Resources, 1986/a; p. 2).

The Siting Process Task Force made "compensation to offset unmitigable impacts and to enhance local benefits" one of its basic principles (Siting Process Task Force, 1987). Another fundamental principle of the siting process was that any community that hosts the waste management facility would have the right to select impact management measures, including mitigation and the amount and type of compensation measures.

The Siting Process Task Force prepared guidelines for impact management during Phase One of the siting process, in consultation with local communities (Siting Task Force, 1989). The guidelines included mitigation measures aimed at avoiding or reducing potential negative impacts; compensation measures related to residual impacts and to equity issues; contingency measures designed to aid in the detection of and timely response to unanticipated effects of the facility; and community relations measures aimed at maintaining a positive relationship with the host community.

The guidelines distinguished between impact-related compensation measures, which aim to offset any unavoidable adverse impacts and leave the community no worse off than it was before the facility was built, and equity-related compensation, which leaves the community better off than it was before in exchange for accepting a facility that will provide benefits for the larger community. Examples of impact-related compensation measures include service subsidies to support any required expansion of services, property tax abatement, property value protection, property buy-out, and replacement of anything that has been destroyed or damaged. Equity-related compensation might include local purchasing and hiring policies, tax subsidies, a tax on the material deposited at the facility, co-use of the facility, and bonus services and facilities (Siting Task Force, 1989).

Compensation and incentives were widely discussed in Phases Two and Three of the site-selection process. Communities that decided to stay in the process concluded that it is possible to contain and transport the wastes without exceeding an acceptable degree of risk, and that potential impacts could be outweighed by possible benefits of the facility, including grants in lieu of taxes, jobs, transportation route improvements, and property value protection. An analysis of the decisions of the communities found that compensation was not a key factor in their decisions (Armour, 1991).

Environmental and socioeconomic impact assessment will take place in Phase Four. The Siting Task Force recommended that in Phase Four negotiations with the host community on impact management also take place and the upper limit of equity compensation be made known to seriously interested communities. It was also suggested that a mutually agreeable impact management package should be defined before a referendum is held (Siting Task Force, 1990/b).[82]

As of December 1992, Phase Four of the siting process was in progress.

ANALYZING THE CANADIAN CASE HISTORY FROM THE PERSPECTIVE OF THE COMPETING VALUES APPROACH

From the consensual perspective -- which emphasizes participatory processes that are intended to develop broad-based support for decisions -- the Canadian siting process is perhaps the strongest of those that we studied, although this was not initially the case. The siting process conducted by Eldorado Resources Ltd. during the early 1980's contained little provision for public participation. This process eventually met with vigorous public protest, and the federal government ordered its cessation in 1986. The current process, proposed by the Siting Process Task Force in 1987, is a highly participative one, however.

The Canadian process gives communities the right to decide whether to accept a proposed facility. They are partners in decision making on all key issues, including disposal method selection, impact management measures, site monitoring and health monitoring. They also have the opportunity to negotiate compensation and benefits with the government.

In potential host communities, liaison groups have been established to work with the Siting Task Force and to conduct public consultation programs. They have been provided training so that they are able to participate effectively in the process. These liaison groups are involved in information gathering and assessment. They also have the power to make recommendations to municipal councils regarding a wide variety of issues, including whether or not to participate, preferred sites, and preferred

---

[82] The original plan (Siting Process Task Force, 1987) suggested that a poll be used to measure community acceptance at the end of Phase Four. The participating communities, however, advised the Siting Task Force that a referendum would be more appropriate (Siting Task Force, 1990/b).

disposal methods. In addition, a number of information sessions have been organized for the general public, and public attitudes have been assessed through opinion polls. At the end of the process, local referenda in the candidate communities are planned.

The high degree of participation appears to have been successful in achieving substantial levels of public support in most potential host communities for the decisions made by the local governments and the Siting Task Force. The recent decision by the municipal council of the Town of Port Hope to become a volunteer community indicates growing confidence in the process.

The Canadian process also rates as a strong one from the political perspective, which emphasizes adaptability and legitimacy. The Siting Process Task Force proposed a highly adaptable, flexible process. No detailed plans were prepared. Communities participating in the process were given the right to influence the siting process, including the agenda and the timing of the various activities, and to select the members of the liaison groups, technical advisors and moderators. The Siting Task Force has cooperated with the communities in an open and responsive manner.

The Task Force also collaborated closely with the responsible federal and provincial agencies to ensure adaptation in the face of a changing regulatory context. Interdepartmental consultative groups were established on both federal and provincial levels to ensure that all relevant requirements are met. Overall, the Canadian process appears to be widely perceived as a legitimate one.

The Canadian process, which now spans more than twenty years, has been less successful from the rational perspective, which emphasizes goal-oriented processes and efficiency in decision making. Efforts to clean up areas contaminated with historic waste date back to the mid-1970's. Beginning in 1982, Eldorado made attempts to find a site for a disposal facility, but, in 1986, the federal government ordered Eldorado to stop this process because of public protest in the potential host communities. An independent government agency was then formed to recommend a process for siting a facility. After completion of the planning phase in 1990, it took over a year before the federal government made the decision to continue to the assessment and implementation phases of the process. Because of this delay, the process lost a large degree of momentum and credibility. According to current plans, the siting process will be concluded in 1995.

The Canadian process has also proved less than distinguished from the empirical perspective, which emphasizes data-based processes and accountability in decision making. This was particularly true for the siting process conducted under the auspices of Eldorado. Although the process carried out by the Siting Task Force is better-documented, collection of technical information concerning potential sites has begun only recently. In efforts to improve the accountability of the process, both the Siting Process Task Force and the Siting Task Force have consulted a number of federal and provincial agencies, private companies, as well as individual experts on regulatory, technical, and socio-economic issues. Communities are also provided funding that allows them to hire technical advisors. Despite the relative lack of emphasis on data-based processes, the Canadian process has not yet been challenged in terms of its degree of accountability.

Chapter 9

FRANCE

BACKGROUND

With 47 nuclear power plants and 70 percent of their electricity generated from these plants, France's nuclear power industry is second in size only to that of the United States (Emel et al., 1990; Kemp, 1989). In addition, France has pursued the development of a full cycle of nuclear fuels production and reprocessing (Emel et al., 1990). The nuclear power industry in France is highly centralized. Electricite de France (EDF), the government-owned electric power provider, is the sole customer for nuclear domestic power plants (Emel et al., 1990). This industry is overseen by Compagnie Generales des Matieres Nucleaires (COGEMA), a subsidiary of the Commissariat a l'Energie Atomique (CEA). The CEA is responsible to the Ministry of Industry, Telecommunications, and Tourism (Emel et al., 1990).

The French distinguish between short-lived and long-lived wastes. Short-lived wastes, which constitute 95 percent of the volume and one percent of the activity, are generally stored to decay (Kemp, 1989). Long-lived wastes are divided into two categories: 1) low- and intermediate-level wastes, and 2) high-level wastes. High level, long-lived wastes are to be disposed of in a deep repository following an interim storage period (Kemp, 1989).

The first LLRW disposal facility, at Centre de la Manche, was opened in 1969 near the La Hague Reprocessing Plant (Parker et al., 1987; Emel et al., 1990). The site was chosen because of its proximity to the reprocessing plant and because geological studies indicated that it was "adequate" (Parker et al., 1987). In reality, the geology is rather complex -- marked by sandstone, schists, and fractures (Parker et al., 1987). In the late 1970s, the CEA determined that a second near-surface disposal site was needed because the original site was rapidly filling up and extension was not feasible (Parker et al., 1987).

A chronological summary of key events related to the LLRW disposal facility siting in France is given in Table 40.

Table 40. CHRONOLOGICAL SUMMARY OF KEY EVENTS IN
FRANCE

Date	Event
1969	The first near-surface disposal facility for LLRW, Centre de la Manche, was opened near the La Hague Reprocessing Plant.
1979	ANDRA (L'Agence Nationale pour la Gestion des Dechets Radioactifs) was created within the Commissariat a l'Energie Atomique (CEA). The Prime Minister approved a site for a short-lived LLRW facility at Saint-Priest-le-Prugne.
1980	Consideration of this site was dropped.
1981	The Supreme Council for Nuclear Safety (CSSN) appointed the Castaing Commission to provide independent technical advice on the waste disposal issue.
1983	The Castaing Commission recommended that two additional disposal facilities should be established and more than one site should be examined before deciding where to locate a disposal facility.
1984	ANDRA decided to construct one additional facility for the disposal of short-lived, low- and medium-level radioactive waste. The Ministry of Industry established requirements regarding the development of a disposal methodology.
1985	Five sites in the Aube region, two sites in Indre, and two sites in Vienne were investigated. ANDRA received authorization to pursue its on-site investigations in the Department of Aube.
1986	The Ministry of Industry established operating procedures to protect against ionizing radiation. Plans for the construction of a new site near Soulaines were submitted by ANDRA to the government. A Public Inquiry regarding the site was held.
1987	The decree on Declaration of Public Utility, allowing ANDRA to purchase the land for the project, was signed by the Prime Minister.
1988	Construction of the facility began.
1989	The facility was licensed.
1992	The facility became operational.

OVERALL APPROACH

On November 7, 1979, ANDRA (L'Agence Nationale pour la Gestion des Dechets Radioactifs) was created within the central French nuclear research authority, Commissariat a l'Energie Atomique (CEA).   ANDRA is responsible for the long-term safe disposal of all categories of radioactive waste (Kemp, 1989).   ANDRA's steering committee is chaired by the Administrator of CEA and includes representatives from EDF and COGEMA. EDF finances 80 percent of the waste disposal activities, with the remainder financed by the national government (Emel et al., 1990).

Originally, ANDRA planned to search for a site with a geology different from the one at Centre de la Manche. Based on advice from the Bureau des Recherches Geologiques et Minieres (BRGM), ANDRA chose a site for the facility at Saint-Priest-le-Prugne because it was near a uranium mine owned by COGEMA that was closing and, therefore, was convenient in terms of landownership and the availability of skilled workforce. CEA gave their support for the site and with little external input, Prime Minister Andre Gireaux decided to approve the site (Kemp, 1989).

Local community opposition was strong, however, and the mayor of Rouen and the local deputy raised questions with the central government about site selection. Protesters objected to the secretiveness behind the site choice and the fact that other potential sites had not been examined (Kemp, 1989). While ANDRA argued that this was the best possible site, local opposition gained independent advice that the local granite was so fractured that it would be unsuitable for such a facility. During the 1980 presidential election, Francoise Mitterand campaigned against locating the disposal facility at this site, and consideration of the site was eventually dropped in May 1980 (Kemp, 1989).

On December 15, 1981, Prime Minister Mauroy recommended establishing local information committees for all major energy-related installations in France. The committees, which are not mandatory, are established by the General Conseil of the relevant department and include municipal officials, environmentalists, and representatives of trade unions and businesses. The committees generally have around 30 members and can seek out information and advice regarding any local energy-related facility. ANDRA pays the costs of the local committee for its sites and provides them with information, and also pays other costs such as independent advice solicited by the committee (Kemp, 1989).

In order to assess and evaluate technical aspects of waste management, treatment, and disposal, a commission was established by ministerial order in 1981. The Supreme Council for Nuclear Safety (CSSN) appointed Professor Raymond Castaing chairman of the Commission (Parker et al., 1987). The Castaing Commission was asked to provide a broad spectrum of independent technical advice on the waste disposal issue (Kemp, 1989).

In 1983, the Castaing Commission recommended that more than one site should be examined before deciding where to locate a radioactive waste disposal facility and that ideally two additional disposal facilities should be established.

On June 19, 1984, the decision to construct one additional facility for the disposal of short-lived, low- and intermediate-level radioactive waste was made. This followed formal governmental approval of ANDRA's national radioactive waste management program (ANDRA, 1991).

INSTITUTIONAL STRUCTURE

Within the central French government, the Ministries of Research, Industry, Health, and Environment oversee nuclear policy regarding waste disposal. These ministries define policy and protection standards, set regulations, approve technical options, and establish necessary controls. CEA engages in research and development surrounding waste technology and helps develop methods to guarantee that protection standards are met. The responsibility for developing site-selection criteria was given to the Castaing Commission at the request of the central government (Kemp, 1989).

ANDRA, part of the CEA, has the responsibility of designing waste disposal facilities and implementing site-selection processes, as well as building and operating radioactive waste disposal facilities. In addition, ANDRA has been charged with safeguarding the disposal of nuclear waste by setting safety specifications and ensuring that waste producers conform to those specifications (Kemp, 1989).

During the siting process, a report from a Public Inquiry must be submitted to the head of the departmental government for approval. A Public Inquiry is an information dissemination and gathering exercise lasting from one to two-and-a-half months. It is an opportunity for interested parties to gain information about the proposal and express their views to the authorities. The majority of the inquiry is devoted to providing the opportunity for written statements to be submitted, although a few days are set aside for oral hearings, when members of the public can come and present their views. The process is informal and emphasis is on public consultation (Kemp, 1989).

Upon approval by COGEMA, the Interministerial Commission for Nuclear Basic Installations must review and authorize the project; final project approval and licensing is the responsibility of the Prime Minister (Kemp, 1989).

SITE SELECTION FOR THE PERMANENT DISPOSAL FACILITY

After the decision was made to develop a new low- and medium-level radioactive waste facility in 1984, ANDRA began the task of selecting a site, assisted by the Castaing Commission and BRGM. The Castaing Commission established criteria for site selection, primarily in relation to geology. Two factors were considered paramount in determining an appropriate type of site: waste isolation and limitation of radioactivity level. To ensure that these goals would be met, the site must: be safe from natural events that might adversely affect the waste isolation system; and possess certain hydrogeologic and geochemical properties that enable the site to mitigate a potential failure of one of the barriers of the waste isolation system by "controlling" the release of radionuclides into the soil of the site (ANDRA, 1991).

These criteria suggested that the site's hydrology should be uncomplicated and that the water outlets should be nearby and clearly identified. The geochemistry of the site should retard radionuclide migration (ANDRA, 1991).

Following recommendations made by the Commission, BRGM and ANDRA developed an ideal-type geological/hydrogeological model for a LLRW repository (Kemp, 1989). According to this model, the site should be built on a thin layer of sand over a thick clay formation. The clay would serve as a tight protection for waters below the site, while the sand would permit easy engineering of the site, as well as natural drainage for any leaks that would come from the concrete-engineered structures containing the waste (Parker et al., 1987).

ANDRA examined the regions of France to determine where these conditions were available and narrowed the search to three Departments. In October 1984, the site- selection phase was begun in the departments of Aube and Indre; sites in the department of Vienne were subsequently considered (ANDRA, 1991). Five sites in the Aube region, two sites in Indre, and two sites in Vienne, were investigated (Kemp, 1989).

ANDRA went to each of the sites during 1985 to undertake geological surveys and test drilling. Between 1984 and 1985, ANDRA also considered 15 volunteer areas, two of which passed initial screening and were investigated further. The site at Neuvey-le-Roi was ruled out because the local mayor ran into strong local opposition, whereas the site at Cholet was dropped after drilling indicated that the hydrogeology was unsuitable (Kemp, 1989).

In July 1985, after submitting a preliminary application file, ANDRA received authorization to pursue its investigations in the department of Aube in order to prepare to submit a formal file for starting the licensing procedure, according to French regulation (ANDRA, 1991).

Several good sites were found close to Soulaines, where the geology was favorable (Parker et al., 1987). The site judged to be most promising consists of a superficial deposit of sand, 25-30 feet thick, covering a cretaceous layer of clay more than 200 feet thick. There is no tectonic activity; the ground is stable, with no danger of landslide; the hydrology is simple and can easily be simulated; underground water outlets are strictly localized, and their capability for dilution is sufficient; geochemical characteristics present high sorption capability for cesium; and the water table cannot reach the bottom of the disposal modules (ANDRA, 1991). Following detailed on-site "qualification works" (characterization), plans for the construction of a new site were submitted in mid-1986 in accordance with the procedural rules governing the construction of new nuclear installations (Emel et al., 1990).

At this time, the public in the immediate locality favored acceptance of the facility. A bit further away (10-20 km from the site), public opposition to the plans developed (Kemp, 1989).

A Public Inquiry was held from September 29 to November 10, 1986. The conclusions were delivered to the Commisaire de la Republique a L'Aube on December 9, 1986. In principal, the site was approved, although two complications emerged. Because the site would require some deforestation, ANDRA was required to provide compensation and replant saplings on a new site. ANDRA readily agreed to this stipulation. It was also suggested that there should be a direct rail link to the site. ANDRA rejected this suggestion, choosing to truck waste from the local rail station (Kemp, 1989).

The decree on Declaration of Public Utility that allowed ANDRA to purchase the land for the project was signed by the Prime Minister on July 22, 1987. The Clearing Permit authorizing ANDRA to cut down trees on the property was signed by the Minister of Agriculture on July 22, 1987. The site was cleared in the spring of 1988 (ANDRA, 1991).

The Construction Permit was obtained on October 11, 1988, and construction began immediately, although the Interministerial Commission for Nuclear Basic Installations did not approve the project until February 9, 1989. The facility was licensed on September 4, 1989 (ANDRA, 1991).

In January 1992, the first shipment of LLRW was received at the Centre de l'Aube disposal facility. The facility will operate for a minimum of 30 years and has a total waste disposal capacity of 1,000,000 cubic meters (Ricaud & Delaunay, 1992).

PERMANENT DISPOSAL METHOD SELECTION

In August 1984, the Ministry of Industry issued a decree, Fundamental Safety Rule No. 1.2 on Quality Assurance, that established requirements for the development of a disposal methodology. On October 2, 1986, Decree 26-1103 was issued establishing operating procedures to protect against ionizing radiation (ANDRA, 1991).

The facility was designed in accordance with the requirements established in these documents. Its design was based on the technical experience gained from 20 years of operations at the Centre de la Manche, which employed earth-mounded concrete bunkers and incorporated the following elements: isolation of waste into a solid concrete structure built above the highest level of the groundwater table; protection of the waste against rainwater using a watertight cover made with artificial and natural material; collection and monitoring of possible waste-infiltrated water; protection of operating structures by mobile hangars until preliminary cover is placed in position over the filled modules; development of remote handling to reduce doses of radiation received by workers; and comprehensive planning for the physical separation of structures in operation from those under construction (ANDRA, 1991).

The goal of the design is to isolate waste from the environment through the institutional control period (300 years). Waste isolation is provided by a combination of methods focusing on the waste packages, the disposal units and the disposal cap, and the water collection system (ANDRA, 1991).

COMPENSATION AND INCENTIVES

ANDRA believed several factors were important in gaining local public acceptance of the facility. They felt that ensuring public safety was paramount. Good information was also critical. After these two factors were accomplished, the natural economic implications (jobs, better services) of establishing a facility become an influential factor in obtaining local public acceptance (Kemp, 1989).

Economic benefits were important to the local population because the Soulaines site is near three small rural villages that have poor services, high out-migration, and an increasingly aging population structure (Kemp, 1989).

ANDRA has hired local people to work at the Soulaines site; local people make up over 90 percent of the operation and management personnel at the site (Emel et al., 1990). Up to 18 months of training has been provided to local people for some jobs. To assist the community, 34 million francs (1987) have been or will be paid to the local community by the Ministry of Industry and ANDRA. In addition, ANDRA will pay a minimum tax of 1.5 million francs per annum during operation to the three villages at the Soulaines site (Kemp, 1989).

Distribution of this money is determined according to criteria established by the local information committee, the local administration, and parliamentary representatives (Kemp, 1989). Local citizens have decided to create a committee of community representatives to share money across the different communities involved and have used the money for schools, libraries, and economic programs (Emel et al., 1990).

To encourage public acceptance, a local information commission (or committee) has been established for the Soulaines site that meets every one to two months. Local people have no direct participation in decisions about the safety of the site; however, the commission serves in an information exchange role (Kemp, 1989).

ANALYZING THE FRENCH CASE HISTORY FROM THE PERSPECTIVE OF THE COMPETING VALUES APPROACH

The French siting process scores comparatively quite highly in terms of the rational perspective -- which emphasizes goal-centeredness and efficiency -- and in terms of the political perspective -- which emphasizes adaptability and legitimacy. It grades out less highly from the empirical perspective -- which emphasizes data and accountability -- and from the consensual perspective -- which emphasizes participation and supportability.

The observed pattern is consistent with the Competing Values approach, which predicts that processes that emphasize the rational perspective are most likely to do so at the expense of the consensual one. Likewise, processes that emphasize the political perspective are most likely to do so at the expense of the empirical one.

With respect to the rational perspective, in the late 1970's, the French government determined that it needed a second disposal facility for ILRW and LLRW. Through creation of a government agency (ANDRA), the government remained focused on this goal despite some early opposition. The siting process in France was reasonably efficient; it took only ten years from the time the Castaing Commission recommended a new facility until the facility was operational.

The French process placed comparatively little weight on the use and acquisition of data. The Castaing Commission established general site selection criteria primarily related to geology that would serve to isolate waste and limit the radioactivity level. An ideal-type geological/hydro-geological model for the repository was developed and sites that matched the ideal-type conditions were sought. The process did not collect or use extensive data from all regions, however. The process was not particularly clear or well-documented about the methods that were used to evaluate regions in terms of the degree to which they matched ideal-type conditions and, later, to select sites within those regions. Accountability was apparently not a highly critical criterion in the design of the French process.

Public participation also does not appear to have been a primary concern of the siting process. After ANDRA chose an initial site -- based on convenience in terms of land ownership and the availability of a skilled workforce -- local opposition forced the site to be dropped from consideration. The public was not formally involved in the selection of the site, however. Rather, ANDRA attempted to gain local public acceptance by ensuring safety, providing information, and emphasizing the natural economic benefits to be realized by the community. While a Local Information Committee was established in the community adjacent to the disposal facility, local residents did not participate directly in decisions about facility safety.

Despite the lack of a highly participatory process, residents of the community nearest the facility supported the decisions. Members of the public residing 10-to-20 kilometers from the site opposed the decision, however, but this opposition was not strong enough to prevent construction of the facility.

Because public participation was not emphasized and few site selection criteria were specified, ANDRA had great flexibility in conducting the siting process. This flexibility enabled ANDRA to consider volunteer areas. It appears that the French siting decision was widely regarded as fair and legitimate.

Chapter 10

THE NETHERLANDS

BACKGROUND

The Netherlands has two nuclear power plants in operation, at Dodewaard and Borssele. Other users of radioactive materials include research centers, industry, and hospitals (Kemp, 1989). In the Netherlands, a distinction is made between high-level radioactive waste (HLRW), intermediate-level radioactive waste (ILRW), and low-level radioactive waste (LLRW). HLRW, consisting primarily of spent fuel and vitrified waste, is sent to reprocessing facilities in France (La Hague) and Britain (Sellafield). ILRW and LLRW, consisting of short-lived, low-radiation-level nuclides, is stored in the Netherlands (Bueno de Mesquita, undated). Approximately 700 cubic meters of ILRW and LLRW are produced in the Netherlands per year. The volume of waste produced by research, industry, and medicine approximately equals the volume produced by the two nuclear power plants (Codee & Vrijen, 1989).

From 1965 to 1982, the Netherlands dumped LLRW and ILRW into the Atlantic Ocean. In the late 1970s and early 1980s, public opposition (expressed through demonstrations blocking shipments) to this method of disposal grew (Brouwer, 1989; Codee & Vrijen, 1987). By 1982, in response to public opposition, government officials decided to end sea dumping of radioactive waste and began to consider alternative methods of disposal (Hermens, undated).

A chronological summary of key events related to the LLRW disposal facility siting in The Netherlands is given in Table 41.

OVERALL APPROACH

Radioactive waste management in the Netherlands has focused on establishing a long-term storage facility, as opposed to a permanent disposal facility, because the relatively small amounts of waste produced do not justify the expense of siting and constructing a permanent disposal facility (Kemp, 1989).

Table 41. CHRONOLOGICAL SUMMARY OF KEY EVENTS IN THE NETHERLANDS

Date	Event
1981	Government officials decided to end sea dumping of radioactive waste and began consideration of alternative methods of disposal.
1982	The Ministry of Housing, Physical Planning, and the Environment decided to store ILRW and LLRW in an industrial building at Velsen, but North Holland provincial authorities prevented the siting. The Central Organization for Radioactive Waste (COVRA) was founded.
1983	An interim storage facility was opened in Zipje.
1984	The Policy Document on Radioactive Waste was approved by Parliament. The Radioactive Waste Storage Facility Site-Selection (LOFRA) Committee was established to help the government identify a small number of sites where LLRW and ILRW could be processed and stored.
1985	The LOFRA Committee recommended three potential sites (in the municipalities of Borssele and Klundert). A Generic Environmental Impact Assessment was conducted.
1986	COVRA chose the site on the grounds of the Provisional Energy Works of Zeeland, near the Borssele nuclear power plant.
1987	COVRA submitted the license application and site-specific environmental impact statement to the Commission for Environmental Impact Assessment.
1988	The Borssele municipal executive approved an alternative site in its jurisdiction.
1989	The new license applications and the Site-Specific Environmental Impact Statement were submitted to the appropriate authorities. The licenses for the new site were granted.
1991	The long-term storage facility became operational.

The first attempts to address the problem of storing ILRW and LLRW were made in 1982. The Ministry of Housing, Physical Planning, and the Environment purchased an old industrial building at Velsen to store the waste. The Ministry expected to have to comply only with planning regulations, as opposed to environmental regulations, when developing the storage facility. When Minister Winsemius announced the plan in December 1982, however, it was greeted with strong public opposition. Local people and the authorities of Velsen refused to accept the storage facility (Brouwer, 1989).

In 1984, the Dutch government issued the Policy Document on Radioactive Waste, which was subsequently approved by Parliament. According to this policy initiative, all kinds and categories of radioactive waste generated in the next 50 to 100 years must be stored in a retrievable way, above-ground, in an engineered structure. The long-term storage facility, together with a centralized treatment facility, must be located at a single industrial site. The policy also mandates that research focusing on the development of final disposal possibilities within the Netherlands or within an international framework must be conducted. Finally, the Central Organization for Radioactive Wastes must be responsible for the storage of all radioactive waste produced (Codee & Vrijen, 1989; Hermens, undated).

INSTITUTIONAL STRUCTURE

The Radioactive Waste Storage Facility Site-Selection (LOFRA) Committee was established in 1984 by the Radiological Protection Directorate of the Ministry of Housing, Physical Planning, and the Environment to help the government identify a small number of sites where LLRW and ILRW could be processed and stored. The five members of the Committee were chosen for their extensive administrative and political experience (Cornelissen, 1987; Hermens, undated; Kemp, 1989).

The Central Organization for Radioactive Waste (COVRA), founded in December 1982, became fully operational in 1984. Although formation of the organization was initiated by the government, COVRA is a private company. The shares of the company are held by the Government of the Netherlands (10 percent); the owners of the nuclear power plant at Borssele (30 percent); the owners of the nuclear power plant at Dodewaard (30 percent); and the Netherlands Energy Research Foundation (30 percent). ECN represents the non-utility company generators of radioactive wastes. All shareholders nominate a representative to serve as a member of COVRA's board of directors (Codee & Vrijen, 1987).

The task of COVRA, as designated by Parliament, is to manage all kinds and categories of radioactive waste produced in the Netherlands. To carry out this task, COVRA collects, transports, treats, stores, and disposes of radioactive wastes. COVRA was responsible for designing, constructing, and operating the long-term storage facility for radioactive waste, licensed by the Ministers of the Environment, Social Services, and Treasury.

COVRA is required to conduct its financial affairs such that all costs associated with storing radioactive wastes are covered by fees paid by the generators. In addition, a provisional fund has been established to cover the costs of future storage and final disposal (Codee & Vrijen, 1987).

The government retains control over COVRA activities through its position on the board of directors and by mandated regulations. Despite the low percentage of shares held by the government, the governmental representative of the board of directors of COVRA has a vote in all major decisions. The government also maintains regulatory authority and control since COVRA must operate within the guidelines of a license issued under the Nuclear Energy Act (Codee & Vrijen, 1987).

SITE SELECTION FOR THE LONG-TERM STORAGE FACILITY

The LOFRA Committee was asked by the government to identify, no later than October 1985, three potential sites for the storage of all types of radioactive waste.   The Committee was required to devote particular attention to the willingness of the local and provincial authorities to cooperate and the necessity of obtaining planning permission. The LOFRA Committee was not explicitly required to consider environmental concerns (Cornelissen, 1987; Hermens, undated; Brouwer, 1989).

The LOFRA Committee evaluated candidate sites according to whether they were located in an industrial region; whether sufficient space (75 acres) to store wastes generated over the next 50 to 100 years was available; and whether the possibility of discharging cooling water and process water existed (Cornelissen, 1987; Kemp, 1989).

In February 1985, the news media publicly identified these sites after the committee had asked for data verification from each site. This led to public outcry (Brouwer, 1989).   Using these criteria, the LOFRA Committee initially identified 20 potential sites. By April 1985, after more detailed informational screening, 12 potential sites were selected.   The LOFRA Committee then explored the willingness of the provincial and municipal authorities to cooperate. Technical and environmental criteria were not used to select a site (Cornelissen, 1987).

One of the potential sites was under the jurisdiction of the Moerdyk Industrial Area Authority in the village of Klundert, an organization with serious financial problems. During the summer of 1985, the media reported that this Industrial Authority was negotiating with the Department of Economic Affairs to sell approximately 75 acres of their land to build a radioactive waste storage facility. Neither the LOFRA Committee nor the Minister of Housing, Physical Planning, and Environment were informed of these negotiations. As a result, two members of the LOFRA Committee resigned, and significant public opposition was generated (Cornelissen, 1987).

In October 1985, the LOFRA Committee submitted a report recommending two potential sites with sufficient available space and willing local authorities. These sites were located in the municipalities of Borssele and Klundert. The Klundert site was the location negotiated by the Moerdyk Industrial Area Authority.   An additional possible site, also located in Borssele, was offered by the Provincial Energy Works of Zeeland, but the LOFRA Committee did not have time to investigate the feasibility of the site due to the lateness of the offer (Cornelissen, 1987; Brouwer, 1989).

According to the Committee's report, the original Borssele site was slightly more favorable than the Klundert site because the local authorities were more willing to host the facility. The Klundert authorities tried to reverse this conclusion by continuing to demonstrate their eagerness to be the host community (Cornelissen, 1987). For the other ten potential sites, the municipalities refused to accept the facility on anti-nuclear and environmental grounds (Brouwer, 1989). On the basis of the LOFRA Committee's recommendation, the Minister of Housing, Physical Planning, and Environment consulted with concerned authorities in Borssele and Klundert.

In 1985, the Environment Ministry initiated an Environmental Impact Assessment process to determine the impact of a radioactive waste storage facility on the environment. COVRA was responsible for conducting this assessment but lacked the necessary expertise. Nucon-Ecoplan, a private consulting firm, was hired to perform the assessment under the direct supervision of COVRA managers. The environmental impact assessment was divided into two stages: the generic assessment was conducted simultaneously with the site-identification process, and site-specific assessments followed site selection (Codee, 1989; Kemp, 1989; Brouwer, 1989).

COVRA completed the site-independent environmental impact assessment in November 1985. The Generic Environmental Impact Statement concluded that the environmental impacts of the facility are so small that environmental factors should not influence the choice of a site. It also found that even if the most severe accident were to occur, no evacuation of people outside the site boundary would be necessary (Bueno de Mesquita, undated).

An independent advisory body, the Commission for the Environmental Impact Assessment, was established to advise the Minister of Housing, Physical Planning, and Environment on the acceptability of the Generic Environmental Impact Statement. The Commission published its report in January 1986. The Commission concluded that the research conducted for the generic environmental impact assessment was insufficient and that analyses of beyond-the-fence effects under normal operation, individual exposure rates, and local environmental implications of radioactive releases into the atmosphere and surface water were inadequate. As a result, the Commission recommended that a site should not be selected until site-specific environmental impact assessments were conducted for the three sites recommended by the LOFRA Committee (Codee, 1989; Kemp, 1989). Although the Minister of Housing, Physical Planning, and the Environment agreed with many of the points in the Commission and concluded that many of the criticisms of the report were valid, he considered it acceptable because COVRA would later be required to conduct a site-specific environmental impact statement, and the missing information could be obtained then (Hermens, undated).

Based on the LOFRA Committee recommendations and the generic environmental impact assessment, the Minister of Housing, Physical Planning, and the Environment decided that there was no reason to treat the storage of radioactive waste differently from any other industrial activity. Furthermore, for all the sites, the local political authorities were willing to host the facility and no special problems were foreseen regarding physical planning or environmental issues. As a result, the Minister decided to leave the final choice to COVRA, requiring only that the decision be made by July 1, 1986 (Cornelissen, 1987; Hermens, undated; Kemp, 1989; Brouwer, 1989).

In June 1986, COVRA chose the site on the grounds of the Provisional Energy Works of Zeeland, near the Borssele nuclear power plant. Parliament was provided information about site selection but did not vote on its acceptability (Cornelissen, 1987; Hermens, undated). The local residents did not protest the selection of Borssele, nor did the prospect of a nuclear waste storage facility being established in the community play a significant role in the municipal council elections in the spring of 1986 (Hermens, undated).

COVRA then had to prepare a license application involving four documents:

* An application for a building permit submitted to the local municipality;
* An application for a permit to release non-radiological waste water into surface waters, as required by the Surface Waters Pollution Act;
* An application for a foundation-, operation-, and release-permit, as required by the Nuclear Energy Act; and
* A site-specific environmental impact statement (Codee, 1988).

The license application described the general purpose of the facility; the safety philosophy adopted in its design; a description of the site and the surrounding area; and a description of the facility design, including technical details about the various buildings and installations that would be constructed on the site. The application also included a safety report detailing the structure of the operating organization; the radiation protection systems; the expected radiation releases during normal operation, as well as during accidents; and a description of the quality assurance system (Codee, 1988).

As part of the license application process, COVRA was required to perform a site-specific environmental impact assessment. In this assessment, all three sites proposed by the LOFRA Committee had to be considered, although the assessment could be concentrated on the favored site at Borssele. The consulting firm of Nucon-Ecoplan was again hired to conduct the assessment under COVRA supervision. Data were collected on population density, local climate, structure of the ground, flora and fauna, surrounding industries, and the frequency of possible external impacts like flooding or plane crashes (Codee, 1989; Hermens, undated).

When COVRA submitted the license application and site-specific environmental impact statement to the Commission for Environmental Impact Assessment in October 1987, two public information hearings were held. These meetings were sparsely attended. Subsequently, when the Commission released its report in February 1988, the Site-Specific Environmental Impact Statement was favorably reviewed, although the government was criticized for failing to provide more information to the public. After coverage of the report by the local news media, some public opposition to the site developed. Some protestors stated that they did not realize how near the site was to the village before the press coverage (Codee, 1989; Hermens, undated).

The protest was mainly directed at the local authorities, and in March 1988, the Borssele Municipal Council passed a motion to look for another site within the municipality. To prevent unnecessary damage to their relationship with the local authorities and the general public, COVRA and the central government authorities decided to wait for further developments before reacting to this initiative. A municipal working group was established to consider potential sites. In May 1988, this group reported that they had found a site as good as, if not better than, the originally proposed site. This site, located in an area under the jurisdiction of the Sloe Port and Industrial Development Board, was approved by the Borssele municipal executive (Hermens, undated).

COVRA then had to submit new license applications and a new Site-Specific Environmental Impact Statement. Because the new site was near the previous one and still in the municipality of Borssele, it was unnecessary to completely redo the previous applications and assessments. COVRA merely had to supplement the previous documents. In response to the public opposition that developed after the first site-specific environmental assessment, six public information meetings were held in Borssele and the neighboring villages. Few people attended these meetings, and little opposition was expressed. However, local residents requested a public opinion poll on the issue several times. Municipal authorities refused either because they did not want to give the impression that the decision would be made by the public or because they feared the outcome would be negative (Codee, 1989; Hermens, undated; Vlek, 1989).

The new license applications and the Site-Specific Environmental Impact Statement were submitted to the appropriate authorities in January 1989. The Environmental Impact Assessment Commission judged the Environmental Impact Statement to be of sufficient quality to facilitate decision making and indicated that the facility should not create any unacceptable environmental risks (Codee, 1987; Codee & Vrijen, 1989; Hermens, undated). License for the new site was granted in the summer of 1989. COVRA began construction after receiving approval, and the long-term storage facility was operational by the end of 1991 (Cornelissen, 1987).

METHOD SELECTION FOR THE LONG-TERM STORAGE FACILITY

The Policy Document on Radioactive Waste approved by Parliament in 1984 specified that all kinds and categories of radioactive waste must be stored in a retrievable way, above-ground, in an engineered structure. When the LOFRA Committee began the site-selection process, however, no detailed design of the facility was available. COVRA was, therefore, required to evaluate the relative advantages of different approaches to managing radioactive waste as part of the generic environmental impact assessment (Hermens, undated; Kemp, 1989).

The generic environmental impact assessment evaluated three alternative approaches to managing radioactive waste. The first alternative, the proposed plan, consisted of high-pressure compaction of solid ILRW and LLRW; incineration of organic liquids and carcasses; solidification of compacted drums and residues with cement; and storage of spent fuel elements in multi-barrier containers. The second alternative, aimed at maximizing volume reduction, included incineration of all types of ILRW and LLRW, and storage of spent fuel elements in water pools. The third alternative, aimed at minimizing potential emissions, included packaging all ILRW and LLRW in cement, and storing spent fuel in transport/storage containers (Bueno de Mesquita, undated).

Each alternative was assessed according to impacts on transportation, surface water, groundwater, soil conditions, air quality, and noise. External radiation and radionuclide discharge under normal and abnormal operating conditions was also considered, as were the impacts on different social groups, procedures for local consultation, and implications of the facility design on eventual disposal options (Kemp, 1989).

Under normal operating conditions, no significant differences in radiation exposure levels at site-boundaries were found among the three alternatives. To keep radiation exposure at acceptable levels, a site where the third alternative, which emphasized emissions minimization, was used would have to be much larger than a site where the other two alternatives were used. Overall, the conclusions of the Generic Environmental Impact Statement suggested that any of the alternatives would be acceptable (Bueno de Mesquita, undated).

COVRA was required, under the Nuclear Energy Act, to provide in the license application detailed specifications of the waste management technologies to be employed at the facility. As designed, the facility includes an exhibition center, a garage, a building for the treatment of ILRW and LLRW, buildings for the storage of ILRW and LLRW, and a building for the treatment and storage of spent fuel and vitrified waste.

The building for the treatment of ILRW and LLRW includes a high-compression compactor, an incinerator for biological wastes, an organic-inorganic liquid separator, a waste water treatment system, an incinerator for organic fluids, shearing and cutting installations, a cementation station, an incinerator for solid wastes, decontamination facilities, and decay-storage areas for short-lived wastes. Each storage building for ILRW and LLRW consists of four modular units, each having the capacity to store 5500 cubic meters of waste. There is enough space at the site to construct five ILRW and LLRW storage buildings over the operational period of the facility. Because of the specific geography in the Netherlands, the storage buildings are designed to withstand major flooding (Codee & Vrijen, 1987; 1989).

COMPENSATION AND INCENTIVES

In 1984, during the Parliamentary debates about developing a siting process for a long-term storage facility, the Environmental Minster recognized the need for some type of compensation for the host community. Government officials did not want to pursue the issue of compensation directly, however, because of its political difficulty. As a result, when selecting potential sites, the LOFRA Committee only had the authority to explore the willingness of local officials to accept a site; they could not negotiate incentives to induce a community to accept the facility. Government officials indicated that if the benefits associated with development were stressed, the need for additional compensation would appear superfluous (Kemp, 1989; Cornelissen, 1987).

When the license application for the Borssele site was made public, the public criticized the local municipal authorities for failing to negotiate with the national authorities for compensation (Cornelissen, 1987). The primary benefit realized by the host community was in selling vacant land in an industrial area at a higher price than the market value (Kemp, 1989).

INTERIM WASTE MANAGEMENT

While the process for siting a long-term storage facility was underway, there was a need to temporarily store LLRW and ILRW. The Netherlands Energy Research Foundation (ECN), in the municipality of Zype, was selected by the central government as the site for the temporary or interim storage facility for ILRW and LLRW. In March 1983, local officials in Zype agreed to establish a temporary storage facility in their municipality if the Ministry of Defense would: curtail military practice operations in the area; make a decision regarding the location of a long-term storage facility for ILRW and LLRW by January 1, 1986; and permit storage of ILRW and LLRW for five years only, with the option of expanding capacity for an additional five years of storage, if necessary (Kemp, 1989). These conditions were accepted, and the interim storage facility was opened in 1983 and continued in operation through 1991.

ANALYZING THE DUTCH CASE HISTORY FROM THE PERSPECTIVE OF THE COMPETING VALUES APPROACH

The Dutch case history differs from the others that we studied because radioactive waste management in the Netherlands focused on the goal of establishing a long-term storage facility, instead of a permanent disposal facility. Considering this difference in overall goal, comparisons between the Dutch experience and those elsewhere should probably be interpreted with substantial caution. Nevertheless, the Dutch siting process appears comparatively successful from the rational perspective, which emphasizes goal-directed processes and efficiency in decision making. The Dutch government and COVRA, the private company responsible for radioactive waste, remained sharply focused on the goal of establishing a facility, even when public opposition developed to the first site that was proposed and even though an interim storage facility was constructed early in the process. Comparativley, the process was quite efficient in terms of the amount of time and resources required to reach a resolution to the LLRW management problem; it took only seven years from initiation of the process until licensing of the facility.

The Dutch process was also comparatively strong from the political perspective, which emphasizes adaptability in decision processes and the perceived legitimacy of the final decision. In the Netherlands, few regulations explicitly governed the site selection process. This gave the siting committee a great deal of flexibility and allowed them to focus on obtaining support of local authorities. Because there were few strict environmental criteria, COVRA was even able to allow local residents to select an alternative site in their community without greatly delaying the process. Despite such flexibility, however, it is not altogether clear whether the general Dutch public perceives the decision as a fair one.

The Dutch process was mixed in terms of the goals and objectives of the consensual perspective -- which values participatory processes and supportability in decisions. The Dutch process initially did not involve the general public directly in the decision making process, although it did give local governments the power to decide whether or not to accept the facility. When the government tried to site a facility at a certain location without investigating other options, public protests played a strong role in preventing the siting of a storage facility there. Later, after COVRA submitted a license application for a site, community residents protested, arguing that they had not previously realized how close the site was to the village.

A municipal working group was then established to identify other potential sites within the community. An alternative site, acceptable to the residents and COVRA, was located. To avoid further public opposition, six information meetings were held after the new site was selected. Local residents requested a public referendum on the issue but local authorities refused to turn decision making authority over to the public.

Because the siting committee solicited support from local authorities and COVRA was willing to accommodate local concerns, local residents supported the decision and did not attempt to prevent the site from being built. While the public ultimately supported the decision, however, local authorities were criticized for failing to negotiate with the authorities for compensation.

Finally, the Dutch siting process was comparatively rather weak in terms of the goals and objectives of the empirical process -- which emphasizes data-based processes and accountability in decisions. The Dutch process was not heavily data-oriented. Technical and environmental criteria were not used to select a site. Potential sites were selected primarily based on the willingness of local authorities to cooperate, location in an industrial region, size of the site, and the possibility of discharging water. Because data were not used in a clear and precise fashion, the basis for accountability was not strong. Documentation concerning why certain sites were eliminated is not readily available, and the decision making process therefore cannot be easily retraced.

Chapter 11

SWEDEN

BACKGROUND

Sweden currently has 12 operating nuclear power plants owned and operated by four utilities, one of which is the government-owned State Power Board (Vattenfall). The competitive electric power industry includes 13 public and private utilities and coordinates many industry efforts through an industrial association, Central Operating Management. The rest of the nuclear power industry is quite concentrated, however, with a single manufacturer of reactors and a single supplier of heavy equipment, both partly owned by the government. No new plants are being developed (Emel et al., 1990).

In Sweden, radioactive waste is classified in three categories: low-level waste, intermediate-level waste, and high-level waste. High-level radioactive wastes (HLRW) are those materials with high heat flux and high initial radiation levels. Intermediate-level radioactive wastes (ILRW) are materials that have long-lived nuclides with low radiation levels; and low-level radioactive wastes (LLRW) include materials that have short half-lives with low radiation levels. Virtually all ILRW and LLRW produced in Sweden are by-products of the nuclear energy industry (Kemp, 1989).

In 1980, a national referendum proposed adoption of a policy, proposed by the Social Democrats, that nuclear power generation should end no later than 2010. In 1988, the Swedish Parliament decided to start phasing out nuclear power in 1995. In June 1991, however, the Parliament overruled the 1988 decision. According to the new decision, there will be no early phase-out of nuclear power. No new date for the start was determined.

A chronological summary of key events related to the siting of a LLRW/ILRW disposal facility in Sweden is presented in Table 42.

Table 42.   CHRONOLOGICAL SUMMARY OF KEY EVENTS IN
SWEDEN

Date	Event
1972	The Swedish Nuclear Fuel and Waste Management Company (SKB) was created by the four utility companies operating nuclear reactors.
1980	In a national referendum, the Swedes decided that nuclear power generation should end no later than 2010.
1983	License for construction of the Swedish Final Repository (SFR) to dispose of ILRW and LLRW was granted. Construction began on the SFR, located near the reactor site at Forsmark.
1984	The Swedish Parliament passed the Act on Nuclear Activities and the Act on Financing of Future Expenses for Spent Nuclear Fuel. OKG Aktiebolag was granted permission from the SSI to construct a shallow land burial facility.
1986	The first disposal effort at the shallow land burial facility took place.
1989	The site license from the Nuclear Power Inspectorate (SKI) was granted. The SFR for short-lived ILRW and LLRW entered operation.
1992	The Swedish Parliament decided that the phase-out of nuclear power plants would not start in 1995 as previously planned.

OVERALL APPROACH

In 1972, the Swedish Nuclear Fuel and Waste Management Company (SKB) was created by the four utility companies operating nuclear reactors. SKB is responsible for planning, developing, constructing, and operating all facilities for the management and disposal of radioactive waste (Kemp, 1989).

In the mid-1970s, a Government Research Group was asked to evaluate approaches to disposing of radioactive waste. In exploring possible alternatives, the Research Group suggested that Sweden should not rely on shallow land burial for three reasons. First, geological resources provide little topsoil and limited clay, leaving few sites for serious consideration. Second, there have been negative experiences worldwide with shallow land burial, especially in the United States. Finally, Sweden has a strong tradition of building in crystalline rock; rock caves have been used to store all kinds of material (Kemp, 1989). The Government Research Group recommended that all radioactive wastes should go into rock caves.

By the end of the 1970s, SKB decided that Sweden should have one disposal facility for LLRW and IRLW and a second for HLRW, rather than developing one facility for all types of waste or a number of facilities for each type of waste. SKB also decided that safety should be based primarily on the inherent characteristics of the repository, not on waste packaging (Kemp, 1989).

In 1980, however, SKB hired a consulting firm to re-examine shallow land burial for LLRW. Their report favored some use of shallow trench burial and was submitted to the National Institute for Radiation Protection (SSI) (Kemp, 1989).

In 1984, the Swedish Parliament passed two laws related to the disposal of radioactive waste. The Act on Nuclear Activities (lag 1984:3) reflects two basic principles guiding development of disposal facilities. First, the producer of the waste bears the technical and financial responsibilities for the safe disposal of nuclear waste products. Second, the State bears the ultimate and long-term responsibility to dispose of nuclear waste in a satisfactory manner for society (Parker et al., 1987).

The second law enacted was the Act on Financing of Future Expenses for Spent Nuclear Fuel Etc. (lag 1984:5). This law decrees that the holder of a license to possess or operate a nuclear power reactor shall defray the costs of safe handling and final safe disposal of spent nuclear fuel from the reactor and radioactive waste deriving from it; ensure safe decommissioning and dismantling of the reactor installation; and conduct the research and development to perform the first two tasks. The reactor owner shall also defray the costs incurred by the State in supplementing the research and development work, administering the fee system, and monitoring and inspecting final repositories (Parker et al., 1987).

License for construction of the Swedish Final Repository (SFR) to dispose of ILRW and LLRW was granted in June 1983. In the fall of 1983, construction began on the SFR with only low levels of public opposition (Parker et al., 1987).

The site license from the Nuclear Power Inspectorate (SKI) was granted early in 1989. The SFR for short-lived ILRW and LLRW near Forsmark started operation in April 1989 (Kemp, 1989).

The repository consists of a rock cavern with different storage chambers located about 50 meters below the seabed. Water depth on the site is five meters (Emel et al., 1990). In total, the SFR will contain 90,000 cubic meters of operational waste. After the year 2000, the SFR may be expanded to receive radioactive decommissioning waste from the Swedish nuclear power plants (Emel et al., 1990). SFR costs are drawn from internal funds within the industry (Kemp, 1989).

In the mid-1980s, OKG Aktiebolag (OKG), a nuclear power company, applied to the SSI for permission to establish a shallow land burial facility at the Oskarshamn reactor site to supplement the SFR. The site is called Hamnerjarden, a small bay on the Baltic coast which already receives cooling water discharge from the Oskarshamn reactors.

Permission to construct the shallow land burial facility was obtained from the SSI in 1984. SSI requires a three-month advance notice of any disposal activity; details on what materials are going into the site, such as the activity levels of bundles, the specific radionuclide content involved, measurements of each bale, and decay calculations; the presence of less than 10 Bq on the whole site at any one time; the surface dose rate of packages must not exceed .5 mSv/hr; for radionuclides of half-life greater than five years, no single package shall exceed 300 kBg/kg; and no persons shall be allowed on the site for 30 to 40 years, and no building or digging can take place in this time period (Kemp, 1989).

Permission was obtained from the local county environmental authorities in 1984. A public meeting was held to inform the local population of the disposal plans. OKG ran the meeting, which was attended by the SSI and community environmental authorities. Twenty to 30 people were present at the meeting, which, as required, was advertised in advance in the local press and by means of pamphlets. The hearing was intended only to inform the public; no strong complaints were received (Kemp, 1989).

The local safety committee, which consists of approximately ten local, appointed politicians and meets four or five times a year, was then informed of the plans for on-site trench burial. They found the plans acceptable and informed the community through the local press. Their chief reason for accepting the proposal was the "institutional control" rule; i.e., that the trench would be within the reactor site and, therefore, inaccessible and of little concern to the public during its operational lifetime. Activity levels will have fallen to an acceptable level by the time the whole site is scheduled to be decommissioned around 2030 (Kemp, 1989).

Wastes are compacted on site; however, no incineration takes place on site. At the trench, the wastes are placed inside temporary walls four meters high. Sand is packed between the material and placed on top of it, with a stabilizing piece of plastic over the top layer of sand. A soil cap seeded with grass is placed over the site. The trench has a simple concrete base with channels leading out to a sampling point before pumped dispersal into the bay. The SSI did not require the sampling; this was done largely to satisfy the environmental department of the local community.

Cost savings for OKG from using on-site trench burial for these wastes, as opposed to sending them to the SFR, are significant. OKG sees this as a method of gaining useful experience with a technique not previously employed by the Swedish nuclear industry (Kemp, 1989).

The first disposal effort at the facility took place in December 1986.

INSTITUTIONAL STRUCTURE

The Swedish Nuclear Fuel and Waste Management Company (SKB), a joint venture of the nuclear utilities, was formed in 1972 (Kemp, 1989). SKB has the responsibility of carrying out a comprehensive policy for spent fuel and waste management. SKB is funded by a fee on nuclear generation; and it is responsible for developing, planning, constructing, and operating facilities for handling and disposing of radioactive waste from the Swedish nuclear power reactors (Parker et al., 1987). SKB hired the Swedish State Power Board, Vattenfall, to design, build, operate, and maintain the SFR (Emel et al., 1990).

The Ministry of Industry oversees the State Power Board and its budget and, thus, has a significant influence on the nuclear power industry, because the State Power Board owns seven of the 12 reactors in Sweden. Nuclear power in Sweden is regulated by the Ministry of Environment and Energy and its administrative subdivisions.

The Nuclear Power Inspectorate (SKI) licenses, supervises, and controls safety in the design, construction, and operation of nuclear facilities. SKI consists of two technical offices: one for regulation and research, and the other for inspection. It also relies on the assistance of consultants from inside, as well as outside Sweden. SKI performed the licensing of the SFR (Emel et al., 1990; Parker et al., 1987).

The National Board for Spent Fuel (SKN) supervises management of spent fuel and waste management funding, and administers the collected capital to cover expenses for spent fuel management and for decommissioning nuclear power stations. SKN is also inspectorate for the research and development carried out by SKB on behalf of the reactor owners (Emel et al., 1990; Parker et al., 1987).

The National Institute for Radiation Protection (SSI) sets and enforces radiation protection standards. Radioactive waste questions are handled within the nuclear power division (Emel et al., 1990; Parker et al., 1987).

SKI, SSI, and SKN each have their own research and development budgets, which they are free to spend as they see fit. To facilitate coordination and exchange of information, a special advisory committee, KASAM, was established in 1985. All three authorities are represented on the advisory committee, which issues an annual status report on nuclear waste handling (Parker et al., 1987).

When constructing a disposal facility, SKB applies for necessary licenses, and several governmental authorities, including SSI and SKN evaluate the applications before decisions are made by SKI. When siting nuclear facilities, Swedish legislation confers the right of the local community to veto the project. The ability to veto is regulated through the Construction Act (Byggnadslagen 1947:385), which states that siting major industrial activities, of considerable importance for the management of energy, or wood, ground, or water resources, shall be decided by the government, and that approval can be given only after community consent (Parker et al., 1987).

While implementation of waste management policies is a responsibility of the waste generators, ultimate responsibility for safe disposal lies with the government. Expert advice and review is provided through the Swedish Consultative Committee for Nuclear Waste Management, which is independent of SKN and reports to the Ministry of Environment and Energy (Emel et al., 1990).

SITE SELECTION FOR THE PERMANENT DISPOSAL FACILITY

Site selection for the SFR was confined to the four reactor sites -- Forsmark, Oskarshamn, Barseback, and Ringhals -- as well as one other area, Studsvik. All sites could provide handling facilities, radiation protection experience, nuclear-related infrastructure, and construction infrastructure. Furthermore, these sites had the required geology (Kemp, 1989).

The SFR was to be placed at a depth of 60 meters. A number of engineering issues related to this depth requirement led to preference for an east coast site, where rock is less fissured (either Oskarshamn or Forsmark). Forsmark was eventually chosen because it had better harbor facilities. This site also had the advantages of readily available water treatment and a labor supply because construction of the Forsmark plant was just ending (Kemp, 1989).

The Forsmark community accepted the proposal. It was expected that 400 jobs would be created at peak SFR construction periods. Approximately 20 to 25 jobs would be created during operation of the facility (Kemp, 1989).

While the application for a license from the SKI was being processed, SKB provided the local population with informational brochures that described the design, emphasized the multiple barriers that would be used to isolate waste from the environment, and made assurances that there would be no environmental consequences. After the SKI licenses were issued and work started, additional information was provided (Kemp, 1989).

In general, the siting procedures for the SFR were performed without much public discussion. Three public meetings were held, one on-site and two in local communities. Arranged by the Local Safety Committee to provide information only, each meeting was attended by approximately 50 people. Although there was some opposition, the general reception was positive (Kemp, 1989).

The People's Campaign Against Nuclear Power, an organization founded in Sweden in 1977, criticized the SFR repository, arguing that disposal of radioactive waste in the repository should be considered sea-dumping because the rock on top of the repository is not completely tight, and sea-dumping is not allowed by Swedish law. The critics also argued that nobody knows to what extent concrete will remain a barrier over a 500-year period and that a major fracture zone extends through the access tunnel regions of the rock.

Neither SKB nor the authorities took part in a debate with the representatives from the People's Campaign (Parker et al., 1987).

SKI and SSI granted operating licenses to the facility in March 1988. The first radioactive waste was placed in SFR in April 1988.

PERMANENT DISPOSAL METHOD SELECTION

The goal in designing the repository was to insure that the radiation dose to human beings residing in the area of the repository would be negligible in relation to natural background radiation. All the storage chambers in the repository have been designed for the specific type of waste to be stored there and the way it is packaged. There are long rock vaults and a high rock cavern in which a concrete silo was built. Two parallel tunnels, each one kilometer long, lead from the surface down to the repository (Emel et al., 1990).

In the licensing procedure for the SFR, SSI was particularly concerned with the radiation protection nearby (public radiation protection) and inside the facility (worker radiation protection). SSI concluded that the repository is sub-optimal from technical and economical points of view, suggesting that the facility was "over-designed" and that large amounts of money could have been saved by using the simplest type of rock caverns for all material. SSI suggests that SKB has included extra precautions to facilitate the licensing procedure as a result of "social pressure." When considering the SFR license application, one member of the Board of SSI made a separate reservation to the decision, claiming that there must be a hidden reason for SKB to construct such an expensive facility which is so much over-designed for its purposes (Parker et al., 1987).

SKB was confronted with other criticisms about overdesigning the SFR. Environmentalists suggested that an over-designed SFR is an indication that the hazard from the facility is greater than anyone is willing to admit and that SKB anticipates placing more highly radioactive material in the SFR than is currently acknowledged. Foreign nuclear industries believe that over-designing sets a precedent that has serious cost implications (Kemp, 1989). In response, SKB emphasized that the most critical issue was the inherent safety of the SFR.

According to SKB, however, one aspect of the SFR design was primarily motivated by public concern. SKB said that monitoring the performance of the facility was convenient and possibly informative, but unnecessary. In their view, the safety of the facility does not depend on monitoring; however, monitoring was judged to be very important in gaining public acceptance (Kemp, 1989).

There is no requirement for retrievability at the SFR. It could be accomplished, however, because of the orderly way the wastes are placed in the silo (Kemp, 1989).

COMPENSATION AND INCENTIVES

The local community has benefited from hosting the SFR disposal facility through taxes paid to the community and the church. These taxes are fixed and relate to the profits of the industry. No additional benefits were negotiated between SKB and the local community (Kemp, 1989).

ANALYZING THE SWEDISH CASE HISTORY FROM THE PERSPECTIVE OF THE COMPETING VALUES APPROACH

From the rational perspective, which emphasizes goal-directed processes and efficiency, the Swedish siting process was among the most successful that we studied. At the end of the 1970's, SKB, the company in charge of managing nuclear fuel and waste, decided that a disposal facility for LLRW and ILRW should be developed. SKB remained resolutely focused on this goal, and a disposal facility was operational within ten years.

The Swedish process was also strong from the political perspective, which emphasizes adaptability and strives to obtain legitimacy for the final decision. This was even more true of the site-selection process than of the method-selection process. Comparatively few explicit criteria were specified for site selection, so this aspect of the process was relatively flexible. There was virtually no local opposition to the site.

Requirements for the disposal method were prescribed more explicitly. SKB was criticized by both environmentalists and the international nuclear community for the design of the facility. Environmentalists argued that the high level of safety features designed into the facility provided evidence that it posed a far greater hazard than its proponents admitted; whereas nuclear interests argued that the facility was over-designed and set a bad precedent that would cause disposal costs to increase unnecessarily elsewhere.

From the empirical perspective, which emphasizes data-based processes intended to help achieve accountability, the Swedish process was comparatively weak. Only the four sites where there was already a nuclear reactor were actively considered. Engineering issues were the primary factor in choosing among the sites. The decision-making process was neither data-based nor well documented.

Finally, the Swedish process was comparatively weak in terms of the goals and objectives of the consensual perspective, which emphasizes participation and broad-based support for the final decision. The siting process for the repository did not allow extensive public involvement. Three public meetings were held, but solely for the purpose of providing information. Despite the lack of extensive public participation mechanisms, however, little public opposition to the facility developed. The local community supported the decision; residents were accustomed to living in the immediate vicinity of a nuclear plant, and the repository was expected to create several jobs, especially during the construction phase.

Chapter 12

SWITZERLAND

BACKGROUND

Switzerland has five light water reactors; 39 percent of their electricity is generated from nuclear power. The Swiss distinguish between low-level (LLRW), intermediate-level (ILRW), and high-level wastes (HLRW). LLRW includes waste materials with short-lived, low-level radionuclides; whereas, ILRW consists generally of materials with short-lived radionuclides, although ILRW materials may contain a limited component of long-lived radionuclides. HLRW includes all material with high-level, short- and long-lived radionuclides (Kemp, 1989).

In 1978, the Swiss federal government approved the changes to the Atomic Act which requires the guarantee of "permanent safe management and final disposal" of radioactive waste as a prerequisite to future development of nuclear energy in Switzerland. In 1980, the Federal Commission for Safety in Nuclear Installations (KSA) and the Nuclear Safety Inspectorate of the Federal Office of Energy (HSK) established the R-21 Guideline, which imposes key safety parameters for radioactive waste disposal in Switzerland. First, a repository must be designed so that it can be sealed within a few years. After a repository is sealed, it must be possible to dispense with safety and surveillance measures. In addition, radionuclides that escape into the biosphere from a sealed repository must not, at any time, lead to individual doses exceeding ten mrem per year (Kemp, 1989; NAGRA, 1991).

In September 1984, the Swiss voted on an initiative to prohibit further construction of nuclear power plants. This initiative was defeated in a nationwide referendum by a ten-percent margin. In 1990, the Swiss rejected another initiative demanding that nuclear power should be completely abandoned while approving another initiative that no new nuclear power permits should be issued for ten years (Oberholzer-Gee, 1993).

A chronological summary of major events related to siting a LLRW disposal facility in Switzerland is given in Table 43.

Table 43. CHRONOLOGICAL SUMMARY OF KEY EVENTS IN SWITZERLAND

Date	Event
1972	The National Co-operative for the Storage of Radioactive Waste (NAGRA) was created by the nuclear power companies and the Swiss Confederation.
1978	The Swiss Federal government approved the Atomic Act, which specifies general policies for radioactive waste management. NAGRA began site-selection efforts for a LLRW/ILRW repository.
1980	The Federal Commission for Safety in Nuclear Installations (KSA) and the Nuclear Safety Inspectorate of the Federal Office of Energy (HSK) established the R-21 Guideline imposing key safety parameters for radioactive waste disposal.
1981	Twenty potential sites for the repository were designated to be subjected to further investigation.
1983	NAGRA submitted applications for exploratory drillings and site characterization of three priority sites and five reserve sites.
1984	In nationwide referendum, the Swiss defeated an initiative to prohibit further construction of nuclear power plants.
1985	NAGRA submitted a proposal, Project Gewahr, to the federal government outlining methods to dispose of LLRW/ILRW. Government permission was granted to allow investigation of the three priority sites by borehole drilling and seismic geophysical surveys.
1988	The Council of Ministers determined that the disposal of short-lived LLRW, as outlined in Project Gewahr, was feasible and safe. NAGRA submitted reports to the federal government on the Phase 1 characterization for two sites, at Oberbauenstock and Piz Pian Grand.
1989	The local community of Wurenlingen agreed to host an interim storage facility. A fourth site, at Wellenberg, was identified.
1990	The federal government decided to postpone decisions on license applications for Phase 2 site characterizations until Phase 1 characterizations were completed for all four sites. Phase 1 site characterization began at the Bois de la Glaive and Wellenberg sites. An application for a general permit to establish an interim storage facility at Wurenlingen was submitted to the government.
1992	ZWILAG was in the process of developing a central interim storage facility at a site in Wurenlingen.

OVERALL APPROACH

The changes in the Atomic Act required that the producers of nuclear waste prove that it can be disposed of safely. In response to the Atomic Act, the Federal Department of Transport, Communication, and Energy (EVED) required that the nuclear power industry conduct a study of the disposal of radioactive wastes as a prerequisite for extending operational licenses beyond 1985. From this study, the government wanted a guarantee (gewahr) that final disposal of radioactive wastes is feasible and safe. In January 1985, the National Co-operative for the Storage of Radioactive Waste (NAGRA) submitted a proposal, Project Gewahr, to the federal government (NAGRA, 1991).

In June 1988, based on several reviews by Swiss nuclear safety authorities and government experts, the Council of Ministers determined that the disposal of short-lived LLRW as outlined in Project Gewahr was feasible and safe. The Ministers further recommended implementing an appropriate repository project (Kemp, 1989; NAGRA, 1991).

INSTITUTIONAL STRUCTURE

According to Swiss law, the producers of nuclear waste are responsible for waste management. Hence, the utility companies involved in nuclear power and the Swiss Confederation, which is directly responsible for waste from medicine, industry, and research, joined in 1972 to form NAGRA (Kemp, 1989; NAGRA, 1991). NAGRA is responsible for the final disposal of all nuclear wastes in Switzerland and is a legally constituted private company consisting of seven equal partners, all producers or future producers of radioactive wastes. The partners include the Swiss Federation, Bern; Bernische Kraftwerke AG, Bern; Kernkraftwerk Gosgen-Daniken AG, Daniken; Kernkraftwerk Leibstadt AG, Leibstadt; Kernkraftwerk Kaiseraugst AG, Kaiseraugst; Nordostschweizerische Kraftwerke AG, Baden; and L'Energie de l'Ouest-Suisse, Laussanne SA (Kemp, 1989; NAGRA, 1991; Parker et al., 1987).

The siting and construction of a repository and all preparatory work is regulated by Federal law. The federal government agencies that serve in a regulatory role include the Federal Interagency Working Group on Nuclear Waste Management (AGNEB), the Federal Commission for Safety in Nuclear Installations (KSA), the Nuclear Safety Inspectorate (HSK) of the Federal Office of Energy (BEW), and the Federal Commission on Nuclear Waste Management (KNE) (NAGRA, 1991). Three types of federal permits must be obtained to construct the facility: a general site permit or license, a construction permit, and an operation license. The general permit must be approved by the Federal Parliament (Oberholzer-Gee, 1993).

Local authorities do not officially have the power to reverse siting decisions made by the federal government. However, there are state zoning and conservation laws that compete with Federal laws. It is not clear whether states can be legally forced to accept the repository (Oberholzer-Gee, 1993). However, NAGRA recognizes that the local authorities of the community and canton (state), as well as the general population, must accept the repository if it is to be built. Without the consensus of the population, there will be long delays, even if a formal federal license has been granted (NAGRA, 1991).

SITE SELECTION FOR THE PERMANENT DISPOSAL FACILITY

NAGRA decided to build two repositories to manage utility waste: a deep geologic repository for high-level waste and a less deep geologic repository for remaining radioactive waste. By government decree, NAGRA also was required to provide disposal for radioactive wastes from medical, research, and other sources (Parker et al., 1987).

NAGRA began site-selection efforts for a LLRW/ILRW repository in 1978, before development of Project Gewahr. In searching for a site, NAGRA was not looking for the "best" possible site, but rather wanted to identify sites offering the required high levels of safety (Kowalski et al., 1986; McCombie, 1989).

A detailed site-selection methodology was employed to choose the best three sites. Originally, 100 potential sites were identified in five geologic formations: anhydride (23 sites); alpine marl and claystone (15 sites); opalinius clay (25 sites); "isolated" formations above the watertable (23 sites); and crystalline rock (14 sites). These sites were then assessed with respect to 28 criteria, as shown in Table 44 (Parker et al., 1987).

Experts at NAGRA then rated each site as "good," "average," "acceptable," or "unacceptable" on each criterion. An additional assessment of non-geologic factors (e.g. proximity to settlements, risk of flooding, conservation) provided further information. In 1981, this procedure resulted in the designation of 20 sites to be subjected to further investigation (Kowalski et al., 1986; Parker et al., 1987).

The 20 preliminary sites were further evaluated in 1982 and 1983. This second round of site selection involved a more detailed study of the existing geological and hydrological data; extended design studies on the construction work; and new, systematic consideration of regional and environmental planning. These studies reduced the number of potential sites to 11 (Kowalski et al., 1986; Parker et al., 1987).

The third round involved assessing of the 11 site areas using eight criteria: landscape; settlements; traffic; water protection; military installations; proprietary considerations; disposal of excavated materials; and the need for special local licenses (Parker et al., 1987).

Further assessments were also made of ecological and hydrogeological aspects, construction techniques, and area planning. The sites were then each judgmentally graded by NAGRA experts on a three-point scale for 12

Table 44. SITING CRITERIA IN SWITZERLAND

Available size of the host rock;	Geometrical forecastability;
Rock mechanical stability;	Complexity of the geology;
Water flow through host rock;	Length of water flow pathways;
Speed of water flow;	Greatest water flow zone;
Biospheric conditions (dilution);	Tectonic stability;
Exogenous change in geology;	Hydrogeology;
Nuclide retention;	Aggressivity of water;
Future raw materials use;	Future water use;
Future underground construction;	Time and cost of siting;
Geological risk;	Time for creation of repository;
Costs for creation of repository;	Environmental impact;
Repository aspects influencing	Traffic;
environment;	Water protection; and
Nature conservatory;	Time needed for permit
Economic significance;	

Source: Parker et al., 1987.

criteria, grouped into the two categories of operations (pre-closure) and nuclear safety (post-closure). Criteria in the operations category included extent of necessary clarification work, construction time, construction cost, environmental compatibility, importance to political economy, and time needed for licensing. Criteria in the nuclear safety category included geology, hydrogeology, biospherical situation, predictability of changes, chemical properties of the geosphere, and prospective use conflicts. The criteria scores were combined, using an unspecified weighting procedure, into aggregate scores for the two broad factors and an overall site desirability score. For the overall score, nuclear safety was given higher priority than development (Parker et al., 1987).

The evaluation produced three priority sites and five reserve sites that were selected for detailed characterization in the municipalities of Bois de la Glaive, Oberbauenstock, and Piz Pian Grand. NAGRA submitted applications for exploratory drilling and site characterization in late 1983 (Kowalski et al., 1986; NAGRA, 1991; Parker et al., 1987).

In September 1985, Government permission was granted for investigation of the three priority sites by borehole drilling and seismic geophysical surveys. In granting permission for the site characterization studies, the federal government required that the individual programs for each site must be divided into sequential phases and that these phases must be executed in parallel at the three sites. Phase One must be limited to geological studies through drilling boreholes while Phase Two would include construction of exploratory tunnels (McCombie, 1989; NAGRA, 1991; Parker et al., 1987).

After getting exploration permits for the three priority sites, NAGRA developed a site characterization plan for a fourth site at Wellenberg, near the Oberbauenstock site (the state government had invited NAGRA to look for possible sites in its jurisdiction). The federal government subsequently granted the exploration permit for this site; however, characterization was not required to proceed in parallel with characterization of the other three sites (McCombie, 1989).

In 1986 and 1987, NAGRA performed Phase One site characterization studies at Oberbauenstock and Piz Pian Grand. NAGRA submitted reports to the federal government on the Phase One characterization for these two sites at the end of 1988. The federal government decided to postpone any decisions on license applications for Phase Two site characterizations until all four preferred sites were characterized. At the Bois de la Glaive site, Phase One site characterization was delayed for several years, not beginning until 1990, due to strong public opposition. Phase One drilling characterizations also did not begin until 1990 at the Wellenberg site. At the Wellenberg site, there has been strong local opposition to drilling an exploratory tunnel for Phase Two site characterization (NAGRA, 1991).

NAGRA has recommended that Phase Two concentrate only on one site or community to save time and money. The federal government insists on parallel research at all four sites in order to be fair and avoid legal prejudgment. As a result, the federal government wants to be granted the right to take land by eminent domain to complete Phase One characterization (Oberholzer-Gee 1993).

PERMANENT DISPOSAL METHOD SELECTION

Because of the high population density in Switzerland, shallow land burial is widely regarded as an infeasible disposal method, even for short-lived LLRW. By 1978, NAGRA was planning to use a horizontally-accessed mine cavern system to dispose of ILRW and LLRW. If a repository is built, the R-21 Guideline established in 1980 by KSA and HSK requires the repository be designed so that it can be sealed within a few years at any time and that safety and surveillance measures can be dispensed with after closing. (Kemp, 1989; NAGRA, 1991; Parker et al., 1987).

In Project Gewahr, a below-ground repository was proposed, and a safety analysis was conducted to determine the feasibility of this approach. Features of the proposed repository design include:

- Underground rock caverns with access through a horizontal tunnel;
- A system of engineered safety barriers including waste solidification matrix (cement, bitumen, polymers), the waste drum, special concrete backfill, concrete linings for the cavern; and
- The sealing of the access tunnel upon closure of the facility.

The repository is designed to ensure supervision and control during operation of the facility that is unnecessary after closure (van Dorp et al., 1986).

COMPENSATION AND INCENTIVES

While there is no federal law requiring compensation and no formal system of compensation or incentives has been developed, utility companies have had a policy of compensating host communities (Oberholzer-Gee, 1993). NAGRA has indicated that it plans to follow a "good neighbor" policy so that local benefits are maximized. While engaged in on-site work at the potential sites, NAGRA assures that workers are housed locally, that local resource taxes are paid, and care is taken to reduce the environmental impacts such as noise associated with drilling. Once a final site is selected, NAGRA will pay local taxes fully, use local contractors whenever possible, and hire locally whenever possible (Kemp, 1989).

Public involvement is encouraged through the use of local Surveillance Commissions at each of the potential sites. These Commissions, consisting of community residents, keep the public informed about the progress of the siting process and inform NAGRA of local needs. When a final site is chosen for the repository, the Commission will serve an important role in communicating community needs and identifying benefits to be negotiated (Kemp, 1989).

INTERIM WASTE MANAGEMENT

As of December 1992, a private company (ZWILAG), founded by the utility companies, was in the process of developing a central interim storage facility at a site in Wurenlingen near the Swiss federal research facility, the Paul Scherrer Institute. In 1989, the local community agreed to host the interim storage facility, and an application for a general permit was submitted to the government in 1990. The general license must be approved by Parliament; thereafter, further applications can be made for construction and operating permits (NAGRA, 1991).

ANALYZING THE SWISS CASE HISTORY FROM THE PERSPECTIVE OF THE COMPETING VALUES APPROACH

From the rational perspective, the Swiss siting process scores highly in terms of being goal-centered, but this focus has not yet led to a resolution of the problem. The goal of the Swiss government in radioactive material management is to ensure that producers of nuclear waste safely dispose of radioactive waste. To achieve this goal, in 1972 the utility companies involved in nuclear power and the group responsible for waste from medicine, industry, and research, established a private company, NAGRA, to be responsible for the final disposal of all radioactive wastes in Switzerland. NAGRA decided to build a repository for all LLRW and ILRW and began site selection efforts in 1978, but a final site has not been selected. Meanwhile, another private company founded by the utility companies began efforts in the late 1980's to establish an interim storage facility.

NAGRA's siting process is strong from the empirical perspective, which emphasizes data-based processes and accountability in decisions. Initially, 100 potential sites were evaluated on 28 technical and environmental criteria. This process resulted in the designation of 20 sites for further investigation, which focused on geological and hydrological data, design studies, and regional planning. When 11 potential sites remained, they were assessed according to eight technical, social, and political criteria. Further evaluation of the potential sites focused on criteria surrounding nuclear safety and operations. This evaluation resulted in three priority sites and five reserve sites that were to be subject to on-site characterization. To date, the process appears to offer a high level of accountability because the information used to make selections as well as the method in which the data were used is clear, apparent, and well documented.

According to the Competing Values theory, it is difficult for processes that are strong from the rational and empirical perspectives also to be strong from the consensual and political perspectives. This is precisely the case for the Swiss siting process.

As for the consensual perspective, which emphasizes participatory processes and supportable decisions, the public has not been directly involved in the site selection process. Public involvement has been encouraged through local Surveillance Commissions; these serve as a means of keeping the public informed about the progress being made on site selection. However, public opposition has greatly delayed site characterization at two of the potential sites. Because a final site has not been selected, it is, of course, unclear whether the final decision will be supported by the local community.

As for the political perspective, which emphasizes adaptability and legitimacy, in Switzerland, the federal government prescribed the site and method selection processes in great detail. As a result, the process has been relatively inflexible and not readily adaptable to changing circumstances. Again, because a final site has not been selected, it is unclear whether the final decision will be considered fair and legitimate by outside observers and the broader body politic.

Chapter 13

FINDINGS AND CONCLUSIONS

This chapter draws some general lessons from the case studies, dividing the major findings and conclusions into the same major issue categories that were used to organize the case studies.[83] Next, we analyze and compare the siting processes from the perspective of the Competing Values framework. Finally, we summarize the most important conclusions and propose recommendations for the design of siting processes.

OVERALL APPROACH

Differences among the various states and countries in their overall approach to the LLRW problem centered on five critical issues: the definition of LLRW; the degree to which LLRW management was an integral part of an overall energy policy; the clarity of the goals of the siting process; the assignment of responsibilities for LLRW management to public versus private organizations; and approaches to public participation.

Definition of LLRW

The definition of LLRW varied considerably among the states and countries that we studied. In the United States, LLRW is defined as radioactive wastes other than irradiated and spent fuel rods from nuclear reactors, high-level radioactive waste from reprocessing spent fuel, transuranic waste, and uranium mill tailings. LLRW is divided into classes A, B, and C, depending on its radionuclide source, half-life, concentration, and energy level.[84]

In Canada, LLRW includes all radioactive wastes excluding irradiated fuel from nuclear reactors and by-products of uranium mining and milling.

---

[83] We did not include interim waste management as a category in this chapter because it was applicable only for a subset of the case studies.

[84] Waste categorized as LLRW may sometimes exceed the concentrations of radioactivity permitted for Class C waste; such LLRW is classified as "Greater than Class C."

LLRW also includes historic wastes that originate from earlier uranium mining activities for which the original generator is unknown.

In addition to LLRW and high-level radioactive waste (HLRW), most European countries classify some waste in a third category as intermediate-level radioactive waste (ILRW). Generally, ILRW consists of materials with relatively low radiation levels, like LLRW, but with longer-lived nuclides. All the European countries in this study dispose of or store ILRW and LLRW in the same facility.

Overall Energy Policy

The various countries and states differed not only in how they defined radioactive waste, but also in the degree to which waste management was an integral part of an overall energy policy that clearly defined the role of nuclear power. In all the places we investigated, the public sees LLRW as strongly related to the nuclear power industry. In most states and countries, the vast majority of LLRW (in terms of radioactivity) does indeed originate from nuclear power plants.

In some countries, overall national energy policies are relatively clear. For example, the Swedish government is committed to phasing out nuclear power production by 2010, whereas Canada and France have expressed strong interest in continued reliance on nuclear energy. Likewise, the Netherlands and Switzerland explicitly decided that continued reliance on nuclear energy was contingent on safe management of radioactive waste. In the United States, however, the role of nuclear energy has not been expressly defined.

Clarity of the Goals of the Siting Process

The various countries and states also differed in the degree of clarity of goals of the siting process itself. In several countries (e.g., Canada, France, the Netherlands, Sweden, and Switzerland), the goals of the LLRW facility siting processes are clear, and the origin, type, and amount of waste to be disposed of are well defined. In other countries, matters are far less clear. For example, in the United States, some states and compacts believe if they establish a LLRW disposal facility, they may have to accept wastes from other states or regions under the emergency provisions of federal legislation. Similarly, Kemp (1992) reported that there were concerns in the United Kingdom that ILRW might eventually be disposed of or stored in a facility originally planned exclusively for LLRW.

Assignment of Responsibilities for LLRW Management to Public versus Private Organizations

The countries that we studied took a variety of approaches to assigning responsibility for the management of LLRW to private versus public organizations. Before 1980, in the United States, generators were responsible for LLRW management. After the federal legislation of 1980, LLRW disposal was made the responsibility of the states, although the states were free to pass along the costs to generators, and all did so. The federal law permits and encourages formation of interstate compacts to develop and manage regional LLRW disposal facilities. These compacts are authorized to exclude out-of-region waste. The federal legislation also required states to take title to and possession of waste and to assume liability for any damages, if facilities for waste disposal did not become available in a timely fashion. In 1992, however, the U.S. Supreme Court found the take-title provision unconstitutional and overturned it.

In Canada, generators are primarily responsible for the management of LLRW. Until a permanent disposal facility is established, LLRW is stored by generators in interim storage facilities. Since 1986, the federal government has assumed responsibility for cleaning up and disposing of historic wastes and is currently attempting to site a facility for disposing of such wastes.

In Sweden and Switzerland, private corporations were formed by major generators for managing radioactive waste. In France, a government bureau in an existing agency was given the same responsibility. In the Netherlands, semi-private corporations, consisting of both generators and government representatives, manage radioactive waste.

Public Participation

In some states and countries, public participation took place primarily through groups and organizations that had already existed before the siting process began. These included, for example, state legislatures, local governments, and existing citizen and environmental groups. In other instances, participation took place through ad hoc groups and organizations that owed their existence, directly or indirectly, to the siting process itself. These included, for example, Citizens' Advisory Committees, Local Liaison Groups, Task Forces, and Local Monitoring or Surveillance Committees.

In most (but not all) instances, there was at least some degree of public participation in each major component of the siting process (viz., defining the overall approach, defining the institutional structure, designing the site- and method-selection processes, selecting a site for the disposal facility, selecting a disposal method, defining the compensation/incentive package, and developing and implementing an interim waste management policy). Typically, public participation was greatest in the site-selection process and less important in more technical areas like the disposal method selection process. Public participation sometimes had an impact even in such technical areas, however. The Nebraska State Legislature passed a law prescribing development of an above-ground structure, and state legislatures in several places, including New York and Texas, prohibited shallow land burial. Public input also influenced the final selection of disposal methods in several other places, including California, Illinois, and Canada.

Across locations, stakeholders' degree of control over the decisions varied considerably and, in a given location, sometimes varied considerably depending on the issue. At least four levels of control were observed. In some instances, stakeholders were involved in the process but had no decision-making power; they received information only. Sometimes, stakeholders were granted the power to review and modify recommendations or decisions. A third level of control gave stakeholders the power to make recommendations, although decision-making power was reserved by state or private agencies or institutions. Finally, the fourth level of control gave stakeholders the direct power to choose a solution or make a decision. Among the states and countries that we studied, the Canadian case probably gave the public the greatest degree of control over decisions.

In each case we studied, state or national legislatures made decisions about the overall approach, the institutional structure, and the type of compensation. In the United States, state legislatures defined the site- and method-selection processes, at least to some extent. In many instances, the relevant legislation specified exclusionary and preference criteria. Sometimes the legislation prescribed the forms of local public participation processes. In some states, legislatures gave guidance concerning the preferred disposal methods (Texas and Nebraska), the preferred siting area (Texas), or the amount of compensation (Texas, Illinois, and Nebraska). In the European countries that we studied, legislative bodies specified broad policy direction but had little if any involvement in specific decisions regarding the siting process. In Canada, Parliament was not involved at all.

In several states, citizen groups, primarily the League of Women Voters, played an important role in citizen education and involvement activities. In California, the League provided a facilitator to the Citizens' Advisory Committee and published educational material for the general public. In New York, the League of Women Voters published educational pamphlets on the LLRW issue and the siting process, and League representatives participated in Citizens' Advisory Committee activities.

In Illinois, Nebraska, and Canada, local governments had the power to decide whether or not they wished to be considered as a potential site of the disposal facility. In Canada, they were also given authority to decide about the disposal method. In Illinois, Texas, Nebraska, and Canada, they were empowered to negotiate compensation, incentives, and community oversight. In the Netherlands, one of the primary siting criteria was the level of support by local governments.

In the states and countries without disposal facilities in place that we studied, generators had to provide for interim waste management when this was required because of delays in siting a disposal facility. In California and New York, generators consequently engaged in direct negotiations with local communities about compensation and incentives. In Sweden, Switzerland, and the Netherlands, generators were directly involved in the entire siting process because they were shareholders in the development companies.

In several states and countries, either citizen and environmental groups or generator organizations played important roles in drafting legislation. In Illinois, environmental groups helped write legislation. In California, waste generators were active in drafting state law. In New York State, local citizen and environmental groups played an important role in changing the state law regarding the overall approach, the institutional structure, and the design of the site- and method-selection processes. Environmental groups played a strong role in determining the nuclear energy policy of Sweden and actively protested against sites in California, Illinois, New York, and Switzerland.

In each investigated state, statewide advisory groups (e.g., Citizens' Advisory Committees) were formed. In most states, these advisory groups included representatives of citizen and environmental groups, generators, and communities near candidate sites. Typically, they had review power over decisions about site and method selection and compensation or incentives. In California, the Citizens' Advisory Committee played an active role in recommending preferred sites.

Local liaison groups (e.g., Local Citizens' Advisory Committees, Community Liaison Groups, Local Monitoring Committees) were formed in California, Illinois, Texas, Nebraska, Canada, Sweden, and France. These groups helped to facilitate public education, elicit public preferences and concerns, advise local governments, and mediate between local residents and the proponents of the siting. In the Netherlands, a local committee proposed an alternative location in their community, which they preferred to the originally proposed site and where the facility was eventually located.

The case studies suggested nine major conclusions about overall approaches to LLRW management:

1. Although the countries differed substantially in how they defined LLRW, such differences did not have a significant impact on the characteristics or outcomes of siting processes.

Some countries, such as the United States and Canada, define LLRW by exclusion. They specify the forms of waste that are classified as HLRW; the remainder is classified as LLRW. Most European countries define LLRW by expressly specifying its characteristics. The United Kingdom, for example, defines LLRW in terms of maximum levels of alpha and beta-gamma concentrations (Kemp, 1992). The United States and Canada define only two major categories of radioactive waste and do not distinguish a category of ILRW. Some long-lived wastes, that would be classified as ILRW in most European countries, are classified as LLRW in the United States and Canada.

2. When LLRW management was part of a broader nuclear energy policy framework, siting processes were more likely to be supported by the public.

In Canada, France, the Netherlands, and Sweden, which have relatively clear energy policies, there appeared to be widespread support for siting processes. As a counter-example, however, siting processes were not widely supported in Switzerland, despite the fact that government policies on nuclear energy are comparatively quite clear. In the United States, where overall energy policy is less clear and the potential cost of eliminating sources of LLRW (e.g., increased energy costs and resource dependence or the potential reduction of some medical treatments) are seldom articulated, the public was much less supportive.

In the United States, some difficulties appeared to stem less from lack of clarity in the overall policy than from many stakeholders' perception that they are unable to participate meaningfully in decision making about fundamental questions of overall policy. A number of opposition groups objected that the overall approach took for granted the need for LLRW disposal facilities, without allowing discussion of what was the key issue in their view -- the future status of nuclear power in the United States. Persons who vehemently oppose nuclear power are unlikely to be satisfied with a clearer energy policy if that policy features a continuing role for nuclear power, or with a greater voice in the siting process if that process implicitly assumes that a facility is needed.

3. In countries where the source, type, and amount of LLRW were clear, there was more public support than in countries where there was ambiguity about the specific waste streams to be disposed of at the facility.

In Canada, France, the Netherlands, and Sweden, where there was relative certainty about the source, type, and amount of LLRW, there was more public support than in countries where there was more uncertainty about these issues. In the United States, potential host communities were concerned about the possibility of later being forced to accept LLRW from other regions. Similarly, Kemp (1992) reported that potential host communities in the United Kingdom were concerned that the proposed facility would eventually be forced to accept ILRW.

4. The public appears to be generally most accepting of an overall management approach when generators are responsible for LLRW disposal. Holding generators responsible may lead to less efficient and accountable processes, however, because responsible parties may not always be identifiable or capable, financially or organizationally, of dealing with disposal responsibilities appropriately.

Before the federal legislation of 1980, generators were responsible for LLRW disposal in the United States. The assignment of the ultimate responsibility for waste disposal to state government was criticized by many environmental and citizen groups as constituting a subsidy to the nuclear industry. In Canada, generators are responsible for the disposal of LLRW, but the national government has had to assume responsibility for historic wastes, for which responsibility is difficult or impossible to assign because the producer may be out of business or unknown. In Sweden and Switzerland, private corporations were formed by groups of generators to manage waste.

5. Making government responsible for LLRW management may help to protect public health and safety and result in a more accountable siting process, but typically results in lower levels of flexibility and efficiency.

State or national governments have been assigned responsibility for LLRW disposal in the United States, Canada, and France. Making government responsible is usually justified on grounds of protecting public health and safety against the consequences of market failures, but this approach places governments in an intermediary role between key stakeholders of the LLRW problem and often results in less flexible and efficient interactions between the parties. The Netherlands and the United Kingdom (Kemp, 1992) have assigned the responsibility for LLRW management to semi-private corporations.

6. Within the United States, strong involvement by state legislatures appears to have increased support and the legitimacy of the decisions.

On the other hand, extensive involvement of the legislature may make decision-making processes less goal-centered and more time-consuming.

In Texas, the State Legislature was involved in all important decisions, even prescribing the siting area legislatively. The involvement of the Legislature led to a process where site selection was re-initiated several times until a consensus was reached.

7. Assigning control over site and method selection and compensation and incentives to local governments increased support for the process. Strong local involvement may make decision processes less goal-centered, however.

In voluntary site-selection processes, there is no guarantee that communities initially interested in the possibility of siting a facility will remain so. This is precisely what occurred in Nebraska. The developer felt the community consented when it initially sent a letter of interest; however, opponents objected that their protests were evidence that the community no longer consented and an affirmative referendum would be necessary before licensing.

8. Granting statewide advisory groups (e.g., Citizens' Advisory Committees) a high degree of control over the site- and method-selection processes is an important factor in enhancing supportability but cannot replace strong involvement of the local government.

Active involvement by the Citizens' Advisory Committee appears to have contributed to making the site-selection process widely accepted in California, at least initially. Failure to respond to the host community's request for compensation and incentives, however, led to erosion of local support.

9. Assigning control over site selection to local governments and citizen groups generally increases support for the process.

In the Netherlands, local governments volunteered to host a facility and, when faced with public opposition, allowed local citizens to identify an alternative, acceptable site in the community. Relinquishing decision-making authority to local citizens enabled a facility to be sited with greater efficiency.

INSTITUTIONAL STRUCTURE

Two basic institutional structures were used in the LLRW disposal facility siting processes in the states and countries that we studied. In some places, including New York, Illinois, Texas, Canada, and France, all major responsibilities were assigned to state agencies. Although they may have hired private contractors for support in planning, constructing, or operating a facility, the designated agencies maintained a high level of control over decisions. In other places, including California and Nebraska, state agencies maintained regulatory control, but private companies were hired as contractors to select a site, construct a facility, and operate it. Similarly, in Sweden, Switzerland, and the Netherlands, private or semi-private organizations were given authority for siting, constructing, and operating the facility, while the government retained regulatory control.

Among states or countries that adopted institutional structures with high levels of state agency control, there was variation in the number and types of agencies involved in the process. The most fragmented structure that we studied was found in New York, where five state agencies shared responsibilities for regulating and licensing LLRW facilities, selecting a site and a disposal method, establishing and operating the facility, and educating and compensating affected communities. Likewise, in Texas, a number of agencies were involved; major tasks associated with LLRW management were shared by several regulatory agencies, one licensing agency, and an ad hoc state agency. In Illinois a single state agency was responsible for licensing and siting LLRW treatment, storage, and disposal facilities, and for providing public education and compensation for local communities. A private company was contracted to select the disposal method and to construct and operate the facility.

Among those states or countries that adopted institutional structures with lower levels of state agency control, there was variation in terms of the degree and type of involvement by private companies. In California, the developer/operator was responsible for site selection and public education, while disposal method selection and compensation remained state responsibilities. In Nebraska, the developer/operator was responsible for proposing and refining the disposal method, as well as selecting the site. While Nebraska law mandated a base level of compensation, the developer had the latitude to negotiate additional compensation.

The case studies suggested four major conclusions about institutional structures:

1. Separating the role of the regulator and the developer enhances the legitimacy of the process.

The approach adopted by Illinois drew criticism from environmental groups because the Legislature assigned the conflicting roles of developer and regulator to the same state agency. Because of serious public concerns about agency self-regulation, Illinois had to change this structure and decrease licensing agency control over site selection.

2. Splitting the responsibilities for site selection, method selection, facility construction, operation, public education, and compensation among different organizations leads to less goal-centered and efficient processes.

New York State's fragmented structure led to uncoordinated efforts in several areas (e.g., public education), while other important areas (e.g., compensation) were left without any real problem owner. Illinois and California, which assigned site- and method-selection responsibilities to different organizations faced similar problems with inefficiency and lack of coordination. In Nebraska, Texas, France, Sweden, and the Netherlands, where site- and method-selection responsibilities were centralized, the siting process was more efficient.

3. Assigning state agencies the responsibilities for site and method selection, facility development and operation, and negotiating compensation will usually result in less adaptable processes because each step in the state agency's decision-making process has to be widely reviewed, discussed, and approved. Private companies (or semi-private organizations) usually have more freedom and flexibility to negotiate settlements with other interested parties.

In New York and Texas, state agencies published detailed site-selection plans in the interests of trying to conduct a highly transparent process. These agencies were later criticized for not following the detailed plans they had developed. In California, Nebraska, Sweden, and the Netherlands, where private or semi-private developers did not publish site-selection plans, site-selection processes have not been as sharply attacked.

4. If private waste management companies invest their own funds, siting processes tend to be more efficient than if processes are financed by governments with fees collected from waste generators. Siting processes conducted by waste management companies, but financed by generators, tend to be less efficient.

In Illinois, California, and Nebraska, which have reached a comparable stage in the siting process, in which license applications have been submitted by the private developer/operator, there are significant differences in cost. In Illinois, the government spent about $85 million on the siting process. In California, the developer/operator company spent about $25 million of its own funds on the process. In Nebraska, where funds collected from generators are invested by the developer company, about $50 million was spent.

DESIGN OF THE SITE- AND METHOD-SELECTION PROCESSES

Since the safety of a disposal facility depends on how a disposal method performs in a particular environment, the purpose of facility siting is to find an appropriate combination of site and disposal methods. In the investigated states and countries, there were three basic approaches to integrating site- and method-selection processes.

In Illinois and Canada, site selection preceded selection of a disposal method. In this approach, preferred sites are identified using a process that considers technical criteria and local acceptance. Next, disposal methods are defined based primarily on the licensability of the facility and local preferences.

In California, Nebraska, the Netherlands, Sweden, France, and Switzerland, selection of a preferred method preceded site selection. This approach emphasizes the importance of the disposal method for the safety of the facility. Before initiating the site-selection process, a preferred disposal method is chosen; then sites that will accommodate this method are sought.

In New York State and Texas, site and method selection were conducted in parallel. This approach restricts the sites and disposal methods under consideration, then follows with a systematic investigation of feasible combinations of potential sites and disposal methods. Another version of this basic approach consists of an iterative process that iteratively and sequentially focuses on various methods and sites.

The case studies suggested five major conclusions about the design of the site- and method-selection processes:

1. The site-first approach is based on the assumption that the performance of the facility should rely primarily on the structure and stability of the site, rather than on facility design.[85] The stringent technical siting requirements of this approach may significantly limit the

---

[85] This principle formed the basis of the NRC 10 CFR Part 61 Regulations.

number of sites considered suitable, especially in states and countries
with less favorable geologic and hydrologic conditions. When potentially
suitable site(s) have been identified, however, method selection can be
more flexible.

> Early in the process, the number of sites in Illinois was narrowed to one
> because of stringent technical requirements and the limitation in the
> authorizing legislation that only voluntary sites could be considered.
> After the preferred site was chosen, the facility designer chose a method
> that considered both specific site characteristics and local preferences.

2. The method-first approach is based on the assumption that the
technology of the facility design is the most significant factor in isolating
waste from the environment. The stringent safety requirements of this
approach may significantly limit the number of alternative methods that
could be considered. When potentially suitable disposal method(s) have
been identified, however, site selection can be more flexible.

> In Nebraska (the eventual host state for the Central Interstate Compact),
> the developer proposed a disposal method before beginning the site-
> selection process. Selection of the proposed technology was based on
> technical considerations, regulatory requirements, and public input. In
> Sweden, France, and the Netherlands, the type of facility to be
> constructed was specified before attempting to locate a site.

3. Parallel investigation of sites and methods is a highly flexible
process, but may become expensive and inefficient.

> New York State designed a parallel site- and method-selection process.
> The intent was to select four candidate sites and five conceptual designs,
> then investigate all feasible site/design combinations. Some members of
> the public sharply criticized this approach for being inefficient and
> creating uncertainty about the type of facility among the potential host
> communities.

4. The method-first approach ensures that the developer can
communicate clearly to the public about the physical characteristics of
the facility. If a method is selected early in the process that is not widely
supported by the public, however, substantial difficulties are likely to
evolve later on.

> California chose shallow land burial as a disposal method, but local
> communities near candidate sites became extremely concerned about the

safety of this technology. As a result of public opposition, various enhancements (e.g., deeper depth trenches, clay-lined sidewalls, and engineered covers) that had been proposed by the local community were adopted.

5. The site-first and parallel processes are problematic because the public cannot be informed about characteristics of the facility, safety regulations, and other important descriptive details until the very last stage of the siting process. The public is unlikely to support a method that the developers cannot clearly describe and defend.

The New York State process was widely criticized because it kept the public uninformed about the disposal technology during the site-selection process. Due to public pressure, the Legislature required that the process be changed so that a disposal method is selected before choosing a site.

In Texas, the number of sites and methods was narrowed down simultaneously. An opinion survey conducted during the site- and method-selection processes indicated that the majority of the people in counties near candidate sites felt that they did not have enough information to make a good judgment about the sites and the methods (TLLRWDA, 1986/b).

SITE-SELECTION FOR THE PERMANENT DISPOSAL FACILITY

The states and countries that we studied took one of two basic approaches to site selection. Some of the states and countries (New York, Texas, Switzerland) adopted processes that were designed to identify the "best available" site. Others (California, Nebraska, Illinois, Canada, France, the Netherlands, and Sweden) adopted processes that were designed to identify licensable and politically acceptable sites.

The first type of process identifies an optimal site, usually rank-ordering alternative sites. A set of evaluation criteria is identified, alternative sites are assessed using these criteria, and the assessments are aggregated into an overall numeric score using multi-attribute utility (MAU) models. In New York State, MAU models with 61 evaluation criteria were applied to select candidate areas and potential sites. In Texas, a MAU model with a hierarchy of 17 issues containing 46 criteria was used to select potential sites for detailed characterization. In Switzerland, a similar analytical approach used 28 criteria.

The second type of process tries to find a licensable and politically acceptable site, not necessarily the ideal one. It usually combines procedures for excluding sites that do not meet licensing criteria with procedures for identifying politically acceptable ones. For example, to assure political acceptance, Illinois and Canada adopted entirely voluntary processes. Local communities could withdraw at any time. In the Netherlands, the primary goal of the siting agency was to locate willing communities; agency officials believed that any technical problems could be managed through engineering solutions. In California, to improve the chances of political acceptance, the operator organized a statewide public participation program. Power to recommend candidate sites was granted to a Citizens' Advisory Committee that included representatives of citizen and environmental groups, Native American organizations, and generators.

The case studies suggested four major conclusions about the design of the site- and method-selection processes:

1. Site-selection processes that attempted to identify the optimal site were highly accountable, but less likely to be adaptable, participatory, or result in supportable decisions than were processes that attempted to identify licensable and politically acceptable sites. It is easier to identify, justify, and defend a site as being acceptable than to identify, justify, and defend a site as being optimal.

Processes that attempt to identify the optimal site typically use formal models, as was done in New York State, Texas, and Switzerland. MAU models clarify the strengths and weaknesses of various alternatives.[86] Because they require explicit judgments about the relative importance of the criteria and the relative merits of the various alternatives for each criterion, these models may facilitate discussions among experts and decision makers. When formal models are used, decisions can be retraced. Most of the information used in these models, however, is subjective (e.g., the relative importance of the evaluation criteria) or uncertain (e.g., assessments of the impacts of the alternative sites on the various selection criteria). Because the input data are inevitably subjective and uncertain, the outcomes of MAU models can always be challenged by opponents of the highly ranked sites.

In Texas, a MAU model helped to structure the site-selection process and focus on the most important characteristics of the candidate sites.

---

[86] Since the late 1970s, MAU models have been widely used in environmental decisions (Keeney, 1977; Merkhofer, 1987). With regard to radioactive waste, DOE used such a model while investigating potential high-level radioactive waste sites (Department of Energy, 1986/b; Merkhofer & Keeney, 1987).

Assessments of the geology and hydrology of the candidate sites were changed when detailed site-characterization data became available. These changes affected the rank-order of the candidate sites. Local communities opposed to the sites questioned these data and challenged the integrity of the site-selection process. In New York State, groups from candidate areas raised questions about the quality of the technical data used in computing the performance scores of candidate areas.

2. Generating the site-selection criteria and assessing the impacts of the alternative sites on these criteria are key steps in MAU approaches. Ideally, stakeholders would collect or generate input data, discuss outputs of the model, and arrive at a consensus concerning the best alternative. In practice, however, the interests and judgments of the various stakeholders and technical experts frequently conflict, and questions arise about who should define the evaluation criteria, who should assign their relative weights, and who should assess the impacts of the alternative sites on the criteria.

In New York State, evaluation criteria were defined by technical experts of the New York State LLRW Siting Commission. Representatives of the public were involved in assigning weights to the criteria, but evaluation of the alternatives and the application of the MAU model was left to the experts. Citizen and environmental groups challenged the criteria and the way these criteria were used for narrowing the roster of candidate sites. These groups also objected that voluntary sites were treated differently from other sites, i.e., some voluntary sites were identified as potential sites, whereas sites with similar scores that had not volunteered were excluded from further consideration.

In Texas, the technical experts of the Low-Level Radioactive Waste Disposal Authority and its contractors defined siting issues and subsidiary factors. The same experts rated each site in terms of each factor and assigned importance weights to each factor. An ad hoc group, including technical experts and members of the Citizens' Advisory Panel, assigned importance weights, however.

3. Because of strong public opposition, states or countries using optimal-site procedures may be forced to switch later to acceptable-site processes.

After considerable public protest in communities located near candidate sites, the Texas Legislature dropped the provision of the law that required selecting "the site that appears from the studies to be the most suitable site." This allowed Texas to focus on sites acceptable by the local communities. The New York State LLRW Siting Commission responded to vigorous public opposition in candidate areas by shifting its focus to working with volunteer communities rather than trying to identify technically optimal sites.

4. Site-selection processes that attempt to identify licensable and politically acceptable sites may be challenged on the basis of lack of accountability and fairness to local communities.

In Illinois, environmental groups charged that studying only volunteer communities resulted in a scientifically unjustifiable site-selection process in which data were manipulated in efforts to satisfy technical requirements. In Nebraska, there were charges of political manipulation of the site-selection process. In most states, environmentalists argued that emphasizing political acceptability usually results in imposing the facility on disadvantaged, powerless, rural communities.

PERMANENT DISPOSAL METHOD SELECTION

As with the site-selection process, the states and countries that we studied took one of two basic approaches to method selection. Two states (New York and Texas) adopted processes designed to identify an optimal disposal method. All the other states and countries (California, Illinois, Nebraska, Canada, France, the Netherlands, Sweden, and Switzerland) adopted processes designed to identify licensable and politically acceptable methods.

Method-selection processes that attempt to find technically optimal methods typically use MAU models to evaluate alternative methods, including various technologies relying on below-ground, underground, and above-ground disposal. In New York State, a set of 20 feasible disposal methods were rank-ordered using a MAU model with 23 criteria, including, for example, dose to the public, dose to workers, costs, monitoring, and waste retrievability. In Texas, 11 disposal methods were rank-ordered by a MAU model incorporating 23 criteria classified into 11 issues, including radiation safety, economics and costs, ease of implementation, and impact on local community, among others. Except for criteria weights, all other model inputs were defined by technical experts.

Method-selection processes aimed at finding a licensable and politically acceptable method attempted to reflect public concerns. In most states and countries, the public opposed shallow land burial and favored engineered facilities. In California, the regulatory agency implemented various enhancements to shallow-land burial proposed by the local community. In

Nebraska, the Legislature passed a law prescribing an above-grade, engineered structure. In Illinois, despite safety concerns raised by technical experts, the developer/operator chose an above-grade, engineered structure preferred by environmental groups.

In several states and countries, the public expressed preference for long-term LLRW storage instead of permanent disposal. Storage is distinguished from permanent disposal by its emphasis on retrievability, monitoring, and ongoing institutional control. Proponents of long-term storage argue that most of the waste will decay in a short period and emphasize the importance of waste retrievability because of the superior capability it provides to react to accidents and to respond to anticipated advances in LLRW management technologies that may enable improvements in LLRW facilities. Regulatory agencies in most countries (e.g., the NRC in the United States and the AECB in Canada) oppose this option for safety reasons. Consequently, few states and countries have investigated this option. To assure retrievability, the Netherlands and France use above-grade long-term storage facilities. Canada is also considering establishing long-term storage for historic waste. In response to public concerns about permanent disposal, New York State initiated a study on the feasibility of long-term storage.

In several states and countries, the public demanded that mitigation measures be taken to prevent or reduce potential negative impacts of the facility, that environmental monitoring programs be instituted, and that public control over facility development and operation be implemented. In California, several measures were incorporated in the design to protect the desert tortoise. A monitoring system was designed to prevent water contamination. In addition, the developer/operator planned to conduct emergency response training for the host community.

The case studies suggested three major conclusions about the method-selection process:

1. Method-selection processes that attempted to find an optimal method were accountable, but enjoyed little public support.

> In Texas, 11 disposal technologies were compared and rank-ordered using a MAU model incorporating 31 evaluation criteria. On the basis of the total scores, below-ground modular concrete canisters were selected as the disposal technology. The public, however, favored above-grade disposal.

2. Method-selection processes that tried to find a licensable and politically acceptable disposal method were usually more adaptable and supportable, but led to less accountable decisions.

In Illinois and Nebraska, the public supported the selection of above-grade, engineered disposal. However, this method received comparatively low rankings by technical experts. Several regulators and technical experts questioned the licensability of above-grade facilities.

3. Mitigation, monitoring, and public control are crucial issues in the acceptance of the facility design.

Experiences in various states and countries show that compensation and incentives do not assure public support unless the public feels that the facility is safe and there is sufficient monitoring and public control over its development and operation. Illinois and Canada attempted to negotiate complex impact management packages with local communities, incorporating mitigation, compensation, independent monitoring, and public control.

COMPENSATION AND INCENTIVES

Compensation is defined as repayment for any necessary expenditures or losses associated with siting and operating of the facility. An incentive is more than reimbursement and is a benefit to motivate local communities to accept a facility (National Governors' Association, 1980). Compensation and incentives may be financial or non-financial and can be provided one time or on a continuous basis during the siting, construction, or operation of the facility.

The investigated states and countries took three basic approaches to compensation and incentives. In California, New York, Sweden, and Switzerland, host communities were compensated for the negative impacts of the siting. In Illinois, Nebraska, Texas, Canada, and France, incentives were combined with compensation. The Netherlands relied on industrial development as a natural compensatory mechanism.

Compensating for negative impacts presumes that all impacts can be identified and quantified. The impacts that are compensated vary across states and countries. In California, compensation was provided only for public service costs created by hosting the facility (e.g., incremental burdens on roads, bridges, fire protection, emergency preparation, and monitoring services). In Sweden and Switzerland, compensatory taxes were offered to the community. In New York State, a more comprehensive local assistance package was proposed -- although not yet implemented -- which, in addition to public service payments, included compensation for residential property losses and quality of life impacts (e.g., uncertainty, anxiety, and stigma resulting from the siting), as well as payment in lieu of taxes.

In some states and countries, non-monetary compensation was offered. For example, in New York, Texas, France, and Switzerland, local hiring and purchasing were parts of the proposed local assistance package.

In several states, a special fund was established to facilitate participation of host communities in the siting process (e.g., reviewing documents, performing independent studies). For example, in Illinois, New York and Nebraska, funds were appropriated to counties and municipalities where potential sites were located to reimburse for expenses related to the siting (e.g., education programs, consulting services, and legal services).

Incentives are usually defined through negotiations between the site proponents (government, generators, or facility operators) and the host communities. Examples of monetary incentives include tipping fees, a tax on the material disposed of at the facility, and other fees paid to the local community from the income of the facility. In Texas, the host county was promised ten percent of the gross revenue of the site and at least ten percent of the contribution paid by the state(s) joining in any future compact arrangement.

Non-monetary incentives include services (e.g., schools, hospitals, and free electricity) and facilities (e.g., the complex might include a research or education center). In France, the facility operator helped improve local health, education, and leisure services.

In some cases, a broad set of compensation and incentives was offered to the host communities. In Nebraska, the developer must pay for services (e.g., emergency equipment and training) and compensate nearby landowners for losses in property value. Potential host counties would be reimbursed for expenses related to the siting. In addition, the host county would receive $1 million a year when the facility is operational.

The third approach relied solely on the natural benefits that are a part of any economic development. In the Netherlands, no formal compensation or incentives were offered to the local community. The prime advantage to a community was the sale of costly land and reliance on local labor during construction and operation of the facility.

The investigated cases suggested three major conclusions about compensation and incentives:

1. Under the compensation-only approach, it is extremely difficult to identify and quantify all economic, social, health, and environmental impacts of a disposal facility. Limiting compensation to quantifiable expenses may increase the accountability of the decision, but results in a less flexible and supportable process. When compensation and incentives are negotiated and tailored to the needs and concerns of the local public, flexible, supportable processes are more likely.

California appears to have lost the support of the local community primarily because of the State's unwillingness to respond to the community's request to be compensated for the stigma associated with hosting a LLRW facility. Illinois and Texas, following a more flexible approach to negotiating incentives, were able to find a local community supportive of the facility.

2.   Compensation and incentives may result in a site supported by the immediate host community, but not by more distant communities that receive lesser levels of compensation or benefit.

In Nebraska, the developer found that providing compensation to the community closest to the proposed site did not ensure support from communities farther from the site.   In Illinois, a municipality volunteered to host the site, but the county in which it was located opposed it.

3.   The compensation-and-incentives approach is sometimes criticized as buying local acceptance using bribery. Some argue that this puts the burden of hazardous facilities on less powerful communities, rather than on communities that generate the waste or have technically superior sites.

In Texas, the majority of the waste is produced by nuclear power plants, located primarily in East Texas. In the first round of the site selection, the technically best sites were found in South Texas.   The location identified as the preferred site, however, is in West Texas, where an economically deprived county is willing to accept the waste because of the incentives (Fowler, 1989).   In Illinois and Nebraska, the use of a volunteer process appeared to result in only economically disadvantaged municipalities and counties being considered for potential sites.

EVALUATING THE SITING PROCESSES FROM THE COMPETING VALUES PERSPECTIVE

Various stakeholders have different expectations about what the facility-siting process should be and should attempt to accomplish. As we noted earlier, the Competing Values framework provides a useful theoretical perspective for analyzing the contradictions inherent in facility-siting processes. We have made use of concepts from the Competing Values approach throughout the analysis and in our issue-by-issue review of the case studies, but considering the case studies from the perspective of the eight Competing Values criteria yields some further insights.

Goal-Centered Processes

The rational perspective on decision making emphasizes goal-centered processes. Processes can be regarded as goal-centered to the degree that they remain constantly focused on the primary problem.

On this criterion, the United States appears to have fared more poorly than the other countries that we studied. In the United States, the goals of state governments involved in LLRW disposal facility siting seem to have changed significantly over time. After enactment of the 1985 LLRW Management Policy Act Amendments, many states initiated serious attempts to site a disposal facility by January 1, 1993. As a result of vigorous public opposition in most host states, however, the more instrumental goal of many states became to maintain access to the existing commercial facilities by complying with the letter, if not the spirit, of mandated, interim federal milestones. As it became increasingly clear that few states were making progress toward siting a facility, states that were proceeding fastest appeared to deliberately slow the siting process (i.e., Texas enacted legislation that intentionally delayed licensing their facility until substantially after what was then anticipated to be the earliest feasible date).

As the 1993 milestone approached, states well along in the process became increasingly cautious about being the first to site a new facility, anticipating the possibility of being required to accept waste from across the nation under emergency provisions of the federal law. The siting process appeared to slow in other states in expectation of possible changes in federal law.

Most other countries examined in this study defined their goal as needing to manage LLRW and ILRW, and maintained a clear focus on achieving this goal.

Efficient Decisions

In the rational perspective, the purpose of goal-centered processes is to help to reach efficient decisions. Defining the efficiency of decisions is no simple task, but we have proposed to define efficiency as positively related to the degree of progress made in siting a facility and negatively related to the amount of resources required to achieve such progress.

Outcomes in the United States, where the process was generally less goal-centered, appear to have been less efficient than in most of the other countries we studied, based on the simple observation that no facilities have been established. In contrast, France, Sweden, and the Netherlands have

succeeded in establishing new disposal facilities.   The length of siting
processes ranged from seven years (the Netherlands) to ten years (France and
Sweden).   In Canada and Switzerland, final sites have not been selected.
Among the states that we studied, from 1985 until 1992, license applications
for disposal facilities have been filed by four states -- California, Illinois,
Nebraska, and Texas.   In Illinois, however, the proposed site was rejected;
and in California and Nebraska, the licensing process has been delayed by
substantial public controversy.   As of the beginning of 1993, detailed site
characterization was in progress in Texas, whereas New York State had
retreated to earlier stages of the siting process.

Data-based Processes

The empirical perspective emphasizes making sufficient use of accurate,
pertinent data and information. To assess whether a process is data-based,
the quality and quantity of available information, as well as the ways it was
used in the process, need to be considered. The various states and countries
differed substantially in the degree to which they emphasized the use of
technical and scientific information.

In New York, Texas, and Switzerland, large amounts of technical
information about geology, hydrology, population, and other variables were
collected and analytically integrated into the siting process. Most of the
states and countries that we studied used computerized data bases (e.g.,
geographic information systems, or GIS) to support and justify site selection.
New York State and Texas combined computerized data bases with formal
models (e.g., multi-attribute utility models) to select superior sites and
disposal methods. New York State and Texas also conducted large-scale
analyses of public perceptions and socioeconomic impacts of the facilities.
There was comparatively less emphasis on data collection in California,
Illinois, Nebraska, Canada, France, the Netherlands, and Sweden.

Accountable Decisions

In the empirical perspective, data-based processes help to reach accountable
decisions that are clear, well documented, and easily justifiable. In the states
and countries that we studied, decisions of governmental agencies were
generally comparatively well-documented, while decisions of private
companies were less so.

In New York State, the law required establishing an information system
accessible to the public in order to help to make site selection retraceable. In
many states and countries, in order to increase accountability, the
organizations responsible for site and method selection consulted a number of
government agencies, private firms, and academic institutions. Technical

oversight committees that included independent experts were established in California, Illinois, New York, France, the Netherlands, and Sweden to provide advice and review on site and method selection or licensing.

In a number of states, the organizations responsible for site and method selection were criticized for using poor quality data, misinterpreting the data that were collected, and ignoring additional sources of important information. This occurred both in states that placed relatively greater emphasis on the collection and use of technical information (e.g., New York and Texas) and in states that placed comparatively less emphasis on these functions (e.g., Illinois and Nebraska). In Illinois, the siting process was challenged because the agency involved with site selection made improper changes in a document without the approval of the authors.

Participatory Processes

The consensual perspective emphasizes participation. In a participatory process, the opinions of all the key stakeholders are considered in each phase of decision making. As we discussed earlier, public participation can be accomplished through a variety of mechanisms, including direct participation (public hearings, inquiries, referenda), representation (elected officials, advisory or local monitoring committees, etc.), or opinion sampling (polls and surveys). Such participation may or may not receive external support or facilitation, and the degree of control over the process may vary from essentially none to final decision-making authority.

In the states and countries that we studied, the most typical forms of public participation were public notice and hearings and citizens' advisory committees, with some degree of review power. In Canada, the Netherlands, Illinois, and Nebraska, local governments had the power to make final decisions whether to accept the facility, and local liaison groups played a key role in advising them. In Texas, the State Legislature was the most important vehicle for public participation, making most of the key decisions regarding site and method selection, compensation, and incentives.

Public education programs were organized in all investigated states and countries. Views and concerns of the general public were solicited in various ways, including public meetings and workshops in California, Nebraska, New York, and Texas; public opinion polls in New York, Texas, and Canada; and referenda in Illinois, Nebraska, Canada, Sweden, and Switzerland.

Supportable Decisions

Participatory processes are intended to contribute toward reaching supportable decisions that will be accepted by the key stakeholders affected by decisions.

In the United States, we frequently found divergent interests and opinions among the populations of local communities, adjacent communities and municipalities, states, and regions. In Texas, the host county supported the facility, but a city at a distance of 45 miles from the site strongly opposed it. At locations in Illinois and New York, the local municipal council supported the siting, while the county government was against it. In Nebraska, the issue of community consent has been contentious; the town closest to the site (approximately two miles away) supported the facility, whereas a town ten miles away opposed it. In Canada, France, and Sweden, there appeared to be significant local support for the facilities. In the Netherlands, high levels of local support appeared to be linked to the willingness of the siting authority to address local concerns.

Adaptable Processes

The political perspective encourages adaptability, flexibility, and creativity in approaches to problems. Processes are adaptable if they can be readily changed in response to unexpected events and interventions.

In some of the states and countries that we studied, the legislature or the regulatory agencies prescribed the site- and method-selection procedures, as well as mechanisms for compensation, in considerable detail (e.g., Switzerland, New York State, Texas), thus making these processes less adaptable. There were also states and countries with relatively few regulations regarding site and method selection, compensation and incentives (e.g., California, Illinois, Nebraska, Canada, the Netherlands, Sweden, France). In these states and countries, the siting process was not as rigidly prescribed and developed more flexibility.

Legitimate Decisions

Adaptable processes are intended to lead toward decisions that are legitimate in that they are accepted by the broader public as fair and appropriate, even under changing political circumstances. Adaptable processes helped to reach decisions that seemed to be widely accepted in Canada, France, the Netherlands, and Sweden. In the United States, however, none of the investigated states was able to adequately respond to the changing political circumstances. In California, the proponents of the siting were not prepared to respond to the concerns of the local community in a flexible way, and, as a consequence, they lost the support of the local government, as well as of the broader political community. In Illinois, the siting agency failed to respond to criticism by environmental groups about the quality of the collected data, and this eventually led to the challenge of site legitimacy.

Satisfying Multiple, Competing Criteria

It is clearly difficult, indeed, it is impossible, to satisfy simultaneously all the desirable, competing values of an idealized siting process. Consider the tension between accountability and adaptability. Processes that emphasized accountability (Switzerland, New York State, and Texas) were clearly, and necessarily, less adaptable. Conversely, processes that emphasized adaptability (France, Sweden, and the Netherlands) were less accountable.

In some states and countries, accountability also competed with supportability. States and countries in which siting plans were extensively detailed and decision models were used (Switzerland, New York State, and Texas) were easily attacked by opponents who were able to point out departures from published plans or problems with the data used, even if those deviations had relatively little impact, at least according to supporters of the process. In the United States, processes that focused on supportability (Illinois and Nebraska) were not subject to the same criticism, but were accused of making siting decisions on political grounds.

In the processes that we studied, accountability competed with efficiency. Countries that were able to establish new facilities (France, the Netherlands, and Sweden) and states that were able to file license applications most quickly (California, Illinois, and Nebraska) were the ones that emphasized accountability the least.

In some cases supportability competed with efficiency. Processes that focused on supportability (Illinois, Nebraska, and Canada) were rather expensive and/or subject to long delays.

Our analyses suggested that different siting processes emphasized different criteria. Switzerland, New York State, and Texas emphasized the empirical perspective, focusing on data-use and accountability. Canada, Illinois, and Nebraska emphasized the consensual perspective, focusing on participation and supportability. California emphasized the rational perspective, with primary focus on goal-centeredness and efficiency. France, Sweden, and the Netherlands were able to focus simultaneously on supportability, efficiency, and legitimacy.

In some instances, the relative emphasis shifted over time. In states and countries where facility siting efforts met with significant opposition, changes in emphasis clearly occurred. For example, after several unsuccessful siting attempts, Texas and New York started to put more emphasis on participation and adaptability, while Illinois put more emphasis on the accountability of the siting decisions.

CONCLUSIONS

Among the states and countries that were studied, the major weaknesses of the siting efforts were lack of public support and perceived legitimacy. In a number of instances, these shortcomings ultimately led to inefficient outcomes -- the expenditure of large amounts of public resources and no resolution to the problem. On the basis of the present case studies, we conclude that the prospects for supportable and legitimate decisions are enhanced when the following ten key provisions are satisfied:

1. The siting process is integrated with broader policy regarding energy production and nuclear industry. A consensus needs to be established that the status quo is unacceptable; there is an important problem to be resolved; and the planned facility is the preferred solution to the given problem.

2. The goals of the waste management program are clear. The source, type, and amount of waste to be disposed of at the facility should be well defined, and there need to be guarantees that no additional types and amounts of waste from additional sources will be shipped to the facility.

3. Responsibility for waste management is shared by generators and governments.

4. Responsibilities for site selection, method selection, facility construction, operation, public education, and compensation are assigned to the same organization. Prospects for success appear to be further improved if these responsibilities are assigned to a semi-private or private organization and if the role of the regulator is assigned to a separate public-sector organization.

5. Site and method selection do not occur simultaneously. The disposal method needs to be identified and made widely known before site selection. Alternatively, the site could be selected first, followed by selection of a disposal method preferred by the local community.

6. The goal of the site-selection process is to identify a licensable site with host community support, rather than trying to identify the optimal site. A voluntary process in which communities are allowed to withdraw from consideration at any time further improves the chances for community support.

7. The goal of the method-selection process is to identify a licensable method with host community support, rather than trying to identify the optimal method. Guaranteeing stringent safety standards and providing for mitigation, independent monitoring, and public control over the facility further improve the chances for community support.

8. A tailor-made compensation and incentive package is negotiated with the host community. The negotiations need to address public perceptions, as well as objectively-based measures, of potential impacts.

9. The host community is directly involved in decision making regarding site selection, method selection, and compensation and incentives. Local governments need to act as decision-making bodies, and local liaison groups need to facilitate public education and consultation.

10. The political leaders in both the legislative and executive branches of government display long-term commitment to siting a facility.

REFERENCES

Acres International Ltd. *A Review of Low-Level Radioactive Waste Disposal Technology.* Niagara Falls. 1987.

L'Agence Nationale pour la Gestion des Dechets Radioactifs (ANDRA). *The French Radioactive Waste Management Program.* 1991.

Anderson, D. B. *Surveillance and Maintenance of the West Valley State Licensed Low-Level Radioactive Waste Disposal Area 1983-1987.* 1987.

Anderson, G. *Report to US Ecology on the Low-Level Radioactive Waste Disposal Site Selection Citizens Advisory Committee.* February 1987.

Anderson, G. *Final Report to US Ecology on the Low-Level Radioactive Waste Disposal Site Selection Citizens Advisory Committee.* March 1988.

Anderson, G. *Disposing of Low-Level Radioactive Waste in California: A Guidebook for Citizen Participation.* June 1990.

Armour, A.M. *The Siting of Locally Unwanted Land Uses: Towards a Cooperative Approach.* Oxford: Pergamon Press. 1991.

Ashford, Town of. *Resolution, Town Board.* July 10, 1991.

Atomic Energy Control Board. *Regulatory Objectives, Requirements and Guidelines for the Disposal of Radioactive Waste - Long-Term Aspects.* AECB R-104. Ottawa. June 1987.

Berkovitz, D. Waste Wars: Did Congress "Nuke" State Sovereignty in the Low-Level Radioactive Waste Amendments Policy Act of 1985? *The Harvard Environmental Law Review,* 11, 437-490. 1987.

Berkovitz, D. Waste Wars: It's Not Over Til It's Over. *Radioactive Exchange.* March 31, 1990.

Blackburn, T. W., & Alvarado, R. *Low-Level Radioactive Waste Disposal Activities in Texas.* Unpublished manuscript. 1985.

Brouwer, H. C. G. M. Current Radioactive Waste Management Policy in the Netherlands. In C. Vlek, & G. Cvetkovich (Eds.), *Social Decision Methodology for Technological Projects.* Dordrecht: Kluwer Academic Publishers. 1989.

Brown, H. *The Low-Level Waste Handbook: A User's Guide to the Low-Level Radioactive Waste Policy Amendments Act of 1985.* Washington, DC: Center for Policy Research, National Governors' Association. 1986.

Bueno de Mesquita, K. G. *Long-Term Storage of Radioactive Waste in the Netherlands: The Environmental Impact Report (EIR).* Undated.

California, State of. Ch. 95. 1982.

California, State of, Department of Health Services (DHS). *Report to the Legislature, Low-Level Radioactive Waste.* Preliminary Report. December 1982.

California, State of. Ch. 1177. 1983.

California, State of. Code of Regulations, Title 17, Group 7. 1984.

California, State of. Assembly Bill No. 1000. 1986.

California, State of. Ch. 52. 1987.

California, State of, Department of Health Services & Bureau of Land Management (DHS & BLM). *State of California Indemnity Selection & Low-Level Radioactive Waste Facility.* Draft Environmental Impact Report/Statement. June 1990.

California, State of, Department of Health Services. *Draft Radioactive Material License.* June 1991.

California, State of, Department of Health Services & Bureau of Land Management. *State of California Indemnity Selection & Low-Level Radioactive Waste Facility.* Final Environmental Impact Report/Statement, Vol. 1-3. April 1991.

Canada Department of Energy, Mines and Resources. Mines Minister Releases Federal Policy on Management of Low Level Radioactive Wastes. News Release, 86/67. April 25, 1986/a.

Canada Department of Energy, Mines and Resources. Port Hope Low-Level Waste Disposal Process Redirected. News Release, 86/180. October 10, 1986/b.

Canada Department of Energy, Mines and Resources. Merrithew Announces Independent Task Force to Study New Approach to Waste Site Selection Process. News Release, 86/215. December 11, 1986/c.

Carnes, R., Copenhaver, E., Reed, J., Soderstrom, J., Sorensen, J., Bjornstad, D., & Peelle, E. *Incentives and the Siting of Radioactive Waste Facilities* (ORNL-5880). Oak Ridge National Laboratories, Oak Ridge, Tennessee. 1982. (Cited by van der Pligt, in press).

Central Interstate Low-Level Radioactive Waste Compact Commission. *A Brief History.* 1990.

Central Midwest Interstate Low-Level Radioactive Waste Compact. *Draft Regional Management Plan: Summary Report.* November 1987.

Chem-Nuclear Systems, Inc. *Design Basis Specification.* 1989.

Chem-Nuclear Systems, Inc. *Illinois Low-Level Radioactive Waste Disposal Facility. Executive Summary.* 1991/a.

Citizens Against Radioactive Dumping. *A Seven Position Statement on the Proposed Low-Level Radioactive Waste Disposal Facility.* January 1989.

City of Needles. *Letter to Reuben Junkert*, Project Director, State of California Department of Health Services. June 26, 1991.

Clements, W., Jr. *Letter from the Governor of Texas to Governor Mario Cuomo.* February 6, 1990.

Clements, W. P. *Letter of Certification from the Governor to Kenneth M. Carr, Chairman,* U.S. Nuclear Regulatory Commission. December 12, 1989.

Codee, H. D. K. Licensing of a Central Treatment and Long-Term Storage Facility. *Nuclear and Hazardous Waste Management International Topical Meeting.* September 1988.

Codee, H. D. K. *COVRA's Experience With Environmental Impact Assessments for a Central Treatment and Storage Facility for Radioactive Waste in the Netherlands.* 1989.

Codee, H. D. K., & Vrijen, I. J. Radwaste in the Netherlands. *Waste Management,* Volume 1. 1987.

Codee, H. D. K., & Vrijen, I. J. The Dutch Treatment and Long-Term Storage Facility for Radioactive Wastes. *Proceedings of the Joint International Waste Management Conference,* Volume 1. 1989.

Cornelissen, H. *Long-term Surface Storage of Radioactive Waste in the Netherlands: The Government's Way of Finding a Location.* 1987.

Cortland Citizens Against Radioactive Dumping, Concerned Citizens of Western New York, & Don't Waste New York. *Memorandum to New York State Legislators.* February 1990/a.

Cortland Citizens Against Radioactive Dumping, Concerned Citizens of Western New York, & Don't Waste New York. *Letter to New York State Legislators.* April 1990/b.

Cortland County Low-Level Radioactive Waste Office. *First Annual Report: New York State's Low-Level Radioactive Waste Program - The Cortland County Experience.* September 1990.

Council of the City of Needles. *Resolution, Opposing the Siting of a Proposed Low-Level Radioactive Waste Facility at Ward Valley.* November 18, 1991.

Cramer, E., Tripodes, J., Chenoveth, M., & Behrendt, D. *Public Information Programs: Preparing the Way.* Presentation at the Health Physics Society Annual Meeting. Boston. June 1989.

Cuomo, M. *Letter to Allegany and Cortland County officials and Concerned Citizens.* November 1989/a.

Cuomo, M. *Letter from the Governor to Robert Bernero, Office of Nuclear Material Safety and Safeguards, U.S. Nuclear Regulatory Commission.* December 27, 1989/b.

Cuomo, M. *Letter from the Governor to Hayward Shealy, Department of Health and Environmental Control, State of South Carolina.* December 27, 1989/c.

Cuomo, M. *Letter from the Governor to Jerome Griepentrog, Department of Human Resources, State of Nevada.* December 27, 1989/d.

Cuomo, M. *Letter from the Governor to Elaine Carlin, Department of Ecology, State of Washington.* December 27, 1989/e.

Cuomo, M. *Letter from the Governor to William Newberry, Office of Nuclear Energy, United States Department of Energy.* December 27, 1989/f.

Davis, C. E., & Lester, J. P. Hazardous Waste Politics and the Policy Process. In C. E. Davis & J. P. Lester (Eds.), *Dimensions of Hazardous Waste Politics and Policy* (pp. 1-36). New York: Greenwood Press. 1988.

Deukmajian, G. *Letter from the Governor of California to Governor Mario Cuomo.* December 20, 1989.

Dunlap, R. E. Trends in Public Opinion Toward Environmental Issues: 1965-1990. *Society and Natural Resources,* **4**. 1991.

Easterling, D., & Kunreuther, H. *Finding an Acceptable Policy for Managing Nuclear Waste: A Bottom-Up Approach.* Presentation at the Annual Research Meeting of the Association for Public Policy Analysis and Management, Denver Colorado. October 1992.

Ebasco Services Inc., Envirosphere Company. *Low-Level Radioactive Waste Disposal: Evaluation of Enhancements to Shallow Land Burial.* Final Report. 1988.

Edwards, W., & von Winterfeldt, D. Public Disputes about Risky Technologies. In V.T. Covello, J. Menkes, & J. Mumpower (Eds.), *Risk Evaluation and Management* (pp. 69-92). New York: Plenum. 1986.

Emel, J., Cook, B., Kasperson, R. E., & Renn, O. *Nuclear Waste Management: A Comparative Analysis.* Nevada Nuclear Waste Project. March 1990.

Englander, T., Farago, K., Slovic, P., & Fischhoff, B. A. Comparative Analysis of Risk Perception in Hungary and the United States. *Social Behavior,* **1**. 1986.

English, M. R. *Siting Low-Level Radioactive Waste Disposal Facilities: The Public Policy Dilemma.* New York: Quorum Books. 1992.

Fowler, G. L. *Siting Low-Level Radioactive Waste Disposal Facilities in the United States.* Unpublished manuscript. 1989.

Gershey, E. L., Klein, R. C., Party, E., & Wilkerson, A. *Low-Level Radioactive Waste: From Cradle to Grave.* New York, NY: Van Nostrand Reinhold. 1990.

Governor Orr. *Conditions for Construction of a LLRW Facility in Nebraska.* Undated.

Governor's Program Bill. *An act to amend the environmental conservation law, in relation to the membership and duties of the commission for siting low-level radioactive waste disposal facilities and the advisory committee on siting and disposal method selection for permanent disposal facilities and repealing paragraph b of subdivision 2 of section 29-0505 of such law relating thereto*, No. 317. 1990.

Griepentrog, J., Shealy, H., & Stanley, R. *Letter to Governor Mario Cuomo.* March 23, 1990.

Griepentrog, J., Shealy, H., & Stanley, R. *Letter to Governor Mario Cuomo.* January 28, 1991.

Hermens, P. A. H. *The Ministry of Housing, Physical Planning and the Environment's Experience of Environmental Impact Assessment in Relation to the Establishment of a Radioactive Waste Processing and Storage Facility.* Undated.

Illinois, State of. *Low-Level Radioactive Waste Management Act.* Public Act 83-991. December 1983.

Illinois, State of. *Central Midwest Radioactive Waste Compact Act.* Public Act 83-1340. September 1984.

Illinois, State of. *Low-Level Radioactive Waste Management Act Amendments.* Public Act 84-496. October 1985.

Illinois, State of, Department of Nuclear Safety (IDNS). *Interim Management Plan for Low-Level Radioactive Waste in Illinois 1986-1993.* November 1985/a.

Illinois, State of, Department of Nuclear Safety. *Public Participation Plan on Low-Level Radioactive Waste Management in Illinois.* November 1985/b.

Illinois, State of. *Low-Level Radioactive Waste Management Act Amendments.* Public Act 84-244. January 1986.

Illinois, State of, Department of Nuclear Safety. *Standards for Selection of Contractors.* Title 32, Part 605. 1986/a.

Illinois, State of, Department of Nuclear Safety. *Requirements for the Disposal of Low Level Radioactive Waste Away from the Point of Generation.* Title 32, Part 606. 1986/b.

Illinois, State of, Department of Nuclear Safety. *Request for Proposals to Design, Develop, Operate, and Close a Low-Level Radioactive Waste Disposal Facility.* November 1987.

Illinois, State of, Department of Nuclear Safety. *Site Identification Plan and Status Report.* January 1988.

Illinois, State of. *Low-Level Radioactive Waste Management Act Amendments.* Public Act 85-1133. July 1988.

Illinois, State of. *Low-Level Radioactive Waste Management Act Amendments.* Public Act 86-1044. June 1990.

Illinois, State of. *Low-Level Radioactive Waste Management Act Amendments.* Public Act 86-1050. July 1990.

Illinois, State of, Department of Nuclear Safety. *Site Identification Plan and Status Report.* November 1990.

Illinois, State of, Department of Nuclear Safety. *Milestones in Low-Level Radioactive Waste Management*. January 1991.

Interagency Task Force. *Developing a New York State Plan for Interim Management of Low-Level Radioactive Waste*. January 1989/a.

Interagency Task Force. *Recommendations of the Interagency Task Force on Interim Management of Low-Level Radioactive Waste*. November 1989/b.

Kasperson, R. E. *Hazardous Waste Facility Siting: Community, Firm, and Governmental Perspectives*. CENTED Reprint No. 55. Worchester, MA: Clark University. 1986.

Kasperson, R. E., Renn, O., Slovic, P., Brown, H. S., Emel, J., Goble, R., Kasperson, J. X., & Ratick, S. The Social Amplification of Risk: A Conceptual Framework. *Risk Analysis*, **8**, 177-188. 1988.

Keeney, R. L. A Utility Function for Examining Policy Affecting the Salmon in the Stalna River. *Journal of the Fisheries Research Board of Canada*, **34**, 49-63. 1977.

Kemp, R. *Planning and Consultation Procedures for Low-Level Radioactive Waste Disposal: A Comparative Analysis of International Experience* (Research Report No. 1). Norwich, England: Environmental Risk Assessment, University of East Anglia. 1989.

Kemp, R. Why Not in My Backyard? A Radical Interpretation of Public Opposition to the Deep Disposal of Radioactive Waste in the United Kingdom. *Environment and Planning*, **22**, 1239-1258. 1990.

Kemp, R. *The Politics of Radioative Waste Disposal*. Manchester: Manchester University Press. 1992.

Kemp, R., & O'Riordan, T. Planning for Radioactive Waste Disposal: Some Central Considerations. *Land Use Policy*, 37-44. January 1988.

Keown, C. F. Risk Perceptions of Hong Kongese vs. Americans. *Risk Analysis*, **9**, No. 3. 1989.

Keystone Center. *Siting Waste Management Facilities in the Galveston Bay Area: A New Approach*. Keystone, Colorado: The Keystone Center. 1982.

Keystone Siting Process Group. *The Keystone Siting Process Handbook: A New Approach to Siting Hazardous Waste Management Facilities*. Austin: Texas Department of Water Resources. 1984.

Kleinhesselink, R. R., & Rosa, E. A. Cognitive Representation of Risk Perception. *Journal of Cross-Cultural Psychology*, **22**. 1991.

Kowalski, E., Gassner, R., & Kappeler, S. Constructional Concepts and Site Selection Procedures for LLW/ILW Repository in Switzerland. *International Symposium on the Siting, Design, and Construction of Underground Repositories for Radioactive Wastes*. March 1986.

Kunreuther, H., Aarts, T. D., & Fitzgerald, K. *Siting Noxious Facilities: A Test of the Facility Siting Credo*. Report #92-03-01. Risk and Decision

Processes Center, The Wharton School of the University of Pennsylvania. March 1992.

Kunreuther, H., & Linnerooth, J. *Risk Analysis and Decision Processes: The Siting of Liquified Energy Gas in Four Countries.* New York: Springer-Verlag. 1983.

League of Women Voters of Nebraska. *Final Report on the Citizens' Advisory Committee on Low-Level Radioactive Waste.* May 1989.

Lentol, J. *Letter from the Assemblyman to Governor Mario Cuomo.* November 30, 1989.

Lewin, A. Y., & Minton, J. W. Determining Organizational Effectiveness: Another Look and an Agenda for Research. *Management Science, 32,* 514-538. 1986.

MacLean, D. Consent and the Justification of Risk Analysis. In V. T. Covello, J. Menkes, & J. Mumpower (Eds.), *Risk Evaluation and Management.* New York: Plenum Press. 1986.

Massam, B. H. *The Right Place: Shared Responsibility and the Location of Public Facilities.* New York: Longman Scientific and Technical. 1993.

Mathews, L. H. *Texas Goes it Alone: Political and Legal Considerations.* Manuscript. 1985.

Matuszek, J. M. Safer Than Sleeping with Your Spouse -- The West Valley Experience. In M. E. Burns (Ed)., *Low-Level Radioactive Waste Regulation.* Chelsea, Michigan: Lewis Publishers, Inc. 1988.

McCombie, C. The Diversity of Waste Disposal Planning in Switzerland. *The Hague Symposium.* October 1989.

McGlennon, J. A. S. The Alberta Experience Hazardous Wastes? -- Maybe in My Backyard. *The Environmental Forum.* February 1983.

Merkhofer, M. W. *Decision Science and Social Risk Management: A Comparative Evaluation of Cost-Benefit Analysis, Decision Analysis, and Other Formal Decision-Aiding Approaches.* Dordrecht: Reidel. 1987.

Merkhofer, M. W., & Keeney, R. L. A Multiattribute Utility Analysis of Alternative Sites for the Disposal of Nuclear Waste. *Risk Analysis, 7.* 1987.

Miller, G., Boxer, B., Cranston, A., Waxman, H., Levine, M., Brown, G., Lantos, T., Torres, E., Pelosi, N., Beinlenson, A., Berman, H., Stark, F., Lehman, R., Banetta, L., Edwards, D., Mineta, N., Martinez, M., Matsui, R., Dellums, R., & Waters, M. *Letter from Congressmen to Pete Wilson, Governor of California.* November 26, 1991.

Monaco, C. *Comments and Considerations Regarding the Low-Level Radioactive Waste Disposal Facility Siting Process.* November 15, 1989.

Monaco, C. *Letter to Thomas Combs, U.S. Nuclear Regulatory Commission.* May 10, 1990/a.

Monaco, C. *Letter to Joseph Colman, U.S. Department of Energy.* May 19, 1990/b.

Monaco, C. *Summary of Points Raised by Governor Cuomo at the May 23, 1990 Meeting.* June 1, 1990/c.

Monaco, C. *Letter to Paula Van Meter, Assistant Counsel to the Governor.* June 7, 1990/d.

Monaco, C. *Challenge of the Federal 1985 LLRW Policy Amendments Act (Part 1): LLRW Article Number 17.* August 6, 1990/e.

Monaco, C. *Challenge of the Federal 1985 LLRW Policy Amendments Act (Part 2).* August 13, 1990/f.

Monaco, C. *Letter to Thomas McGiff, Office of Environmental Health, Cornell University.* January 8, 1991/a.

Monaco, C. *Letter to Robert Bernero, U.S. Nuclear Regulatory Commission.* February 8, 1991/b.

Monaco, C. *Letter to Steven Gavitt, New York State Department of Health.* September 30, 1991/c.

Morell, D. Siting and the Politics of Equity. *Hazardous Waste,* 1. 555-571. 1984.

Morrison, R. W., & Brown, P. A. *Radioactive Waste Management in Canada. Address to the Uranium Institute.* Sixteenth Annual Symposium, London. September 4-6, 1991.

National Co-operative for the Storage of Radioactive Waste (NAGRA). *Nuclear Waste Management in Switzerland.* August 1991.

National Co-operative for the Storage of Radioactive Waste. *Nuclear Waste Management.* Undated.

National Governor's Association. *Low-Level Waste: A Program for Action.* August 1980.

Nebraska, State of. Legislative Bill 200. 1983.

Nebraska, State of. Legislative Bill 491. 1986.

Nebraska, State of. Legislative Bill 426. 1987.

Nebraska, State of. Legislative Bill 1092. 1988.

Nebraska, State of. Legislative Bill 761. 1989.

Nebraska, State of. *Low-Level Radioactive Waste Disposal Act.* 1990/a.

Nebraska, State of. *Central Interstate Low-Level Radioactive Waste Compact Act.* 1990/b.

Nebraska, State of. Legislative Bill 837. 1991/a.

Nebraska, State of. Legislative Bill 838. 1991/b.

Nebraska, State of, Department of Environmental Control (DEC). *Responsbilities of Groups Involved with LLRW Project.* Undated.

New York State. *Low-Level Radioactive Waste Disposal-Study.* Chapter 978 of the Laws of New York. August 8, 1983.

New York State. *Transportation of Low-Level Radioactive Wastes-Permit.* Chapter 508 of the Laws of New York. July 21, 1986.

New York State. *Low-Level Radioactive Waste Management Act.* Chapter 673 of the Laws of New York. July 26, 1986.

New York State. *Low-Level Radioactive Waste Management Act-Amendments.* Chapter 913 of the Laws of New York. July 1990.

New York State. *Assembly Bill 7227-B.* April 2, 1985.

New York State. *Senate Bill S.3790, Assembly Bill 5743.* March 12, 1987.

New York State. *Senate Bill S.6283-A, Assembly Bill A.8748-A.* June 25, 1991.

New York State Assembly. *Memorandum in Support of Legislation A.8532.* 1989.

New York State Department of Environmental Conservation (NYSDEC). *Draft Environmental Impact Statement for Promulgation of 6 NYCRR Part 382: Regulations for Low-Level Radioactive Waste Disposal Facilities.* July 1987/a.

New York State Department of Environmental Conservation. *Final Environmental Impact Statement for Promulgation of 6 NYCRR Part 382: Regulations for Low-Level Radioactive Waste Disposal Facilities.* December 1987/b.

New York State Department of Environmental Conservation. *Recommendations for State Assistance to Localities Affected by the Siting of a Low-Level Radioactive Waste Management Facility.* April 1987/c.

New York State Department of Environmental Conservation. *New York State Low-Level Radioactive Waste Transportation Report for 1987.* September 1989.

New York State Department of Environmental Conservation. *Draft Generic Environmental Impact Statement for Promulgation of 6NYCRR, Subpart 383-6: Financial Assurance Requirements for Low-Level Radioactive Waste Disposal Facilities.* September 1990.

New York State Department of Environmental Conservation. *Hearing Report in the Matter of Proposed Adoption of Part 383 Subpart 393-6 of Title 6 of the Official Compilation of Codes, Rules and Regulations of the State of New York regarding Financial Assurance Requirements for LLRW Facilities in New York State.* May 1991/a.

New York State Department of Environmental Conservation. *Final Generic Environmental Impact Statement for Promulgation of 6NYCRR, Subpart 383-6: Financial Assurance Requirements for Low-Level Radioactive Waste Disposal Facilities.* December 1991/b.

New York State Department of Environmental Conservation. *Draft Proposed Rulemaking for Low-Level Radioactive Waste Disposal Facilities.* December 1991/c.

New York State Department of Environmental Conservation. *Revised Proposed Rulemaking for Low-Level Radioactive Waste Disposal Facilities.* August 1992.

New York State Department of Health (NYSDOH). *General Information: Low-Level Radioactive Waste in New York State, booklet.* 1989/a.

New York State Department of Health. *Disposal Technologies: Low-Level Radioactive Waste in New York State, booklet.* 1989/b.

New York State Department of Health. *Interoffice Memorandum to M. Heller, Chairman of the LLRW Citizen Advisory Committee.* August 29, 1991/a.

New York State Department of Health. *Draft Scope of Work for Establishing an Independent Scientific and Technical Review Panel to Evaluate New York State's LLRW Siting Process.* June 1991/b.

New York State Department of Health. *Program History.* December 1991/c.

New York State Department of Health. *Scope of Work for Establishing an Independent Scientific and Technical Review Panel to Evaluate New York State's LLRW Siting Process.* June 1992/a.

New York State Department of Health. *Notice to New York State Generators of Low-Level Radioactive Waste.* June 1992/b.

New York State Energy Office (NYSEO). *New York State Low-Level Radioactive Waste Management Study: Draft Report.* January 1984/a.

New York State Energy Office. *New York State Low-Level Radioactive Waste Management Study: Final Report.* April 1984/b.

New York State Energy Research and Development Authority (NYSERDA). *1988 Low-Level Radioactive Waste Status Report.* June 1989/a.

New York State Energy Research and Development Authority. *New York State Plan for Storage, Disposal and Management of Low-Level Radioactive Waste Generated within New York State after 1992.* November 1989/b.

New York State Energy Research and Development Authority. *1989 Low-Level Radioactive Waste Status Report.* June 1990/a.

New York State Energy Research and Development Authority. *Briefing for the Siting Commission.* August 1990/b.

New York State Energy Research and Development Authority. *An Update on the Low-Level Radioactive Waste Management Program in New York State.* December 1990/c.

New York State Energy Research and Development Authority. *1990 Low-Level Radioactive Waste Status Report.* June 1991.

New York State Energy Research and Development Authority. *1991 Low-Level Radioactive Waste Status Report.* June 1992/a.

New York State Energy Research and Development Authority. *Draft Interim Report. LLRW Storage Study.* July 1992/b.

New York State Energy Research and Development Authority. *Draft Design Basis Specification: Centralized Storage Facility for Class A Low-Level Radioactive Waste from Medical and Academic Generators in New York State*. November 1992/c.

New York State Low-Level Radioactive Waste Advisory Committee (NYSLLRWAC). *Mission Statement*. 1988/a.

New York State Low-Level Radioactive Waste Advisory Committee. *Semi-Annual Review of the Activities of the Commission for Siting Low-Level Radioactive Waste Facilities*. August 1988/b.

New York State Low-Level Radioactive Waste Advisory Committee. *Second Semi-Annual Report on the Plans and Schedule of the Commission for Siting Low-Level Radioactive Waste Facilities*. March 1989/a.

New York State Low-Level Radioactive Waste Advisory Committee. *Public Meeting Comments*. March 22, 1989/b.

New York State Low-Level Radioactive Waste Advisory Committee. *Third Semi-Annual Report on the Plans and Schedule of the Commission for Siting Low Level Radioactive Waste Facilities*. September 1989/c.

New York State Low-Level Radioactive Waste Siting Commission (NYSLLRWSC). *Request for Proposals for Site Selection and Disposal Method Selection for the Disposal of Low-Level Radioactive Waste in the State of New York, Request for Proposals No. 001-87*. 1987.

New York State Low-Level Radioactive Waste Siting Commission. *Draft Procedure for Selecting Sites for Disposal of Low-Level Radioactive Wastes*. July 1988/a.

New York State Low-Level Radioactive Waste Siting Commission. *Draft Procedure for Selecting Methods for Disposal of Low-Level Radioactive Wastes*. July 1988/b.

New York State Low-Level Radioactive Waste Siting Commission. *Generic Scope of Draft Environmental Impact Statement*. September 1988/c.

New York State Low-Level Radioactive Waste Siting Commission. *Draft Statewide Exclusionary Screening Report*. September 1988/d.

New York State Low-Level Radioactive Waste Siting Commission. *Summary of Site and Method Selection Workshop*. October 1988/e.

New York State Low-Level Radioactive Waste Siting Commission. *Public Meeting Summary Report*. October 1988/f.

New York State Low-Level Radioactive Waste Siting Commission. *Plan for Selecting Methods for Disposal of Low-Level Radioactive Wastes*. November 1988/g.

New York State Low-Level Radioactive Waste Siting Commission. *Plan for Selecting Sites for Disposal of Low-Level Radioactive Wastes*. November 1988/h.

New York State Low-Level Radioactive Waste Siting Commission. *Candidate Area Identification Report*. December 1988/i.

New York State Low-Level Radioactive Waste Siting Commission. *LLRW Frontline*. Winter 1988/j.

New York State Low-Level Radioactive Waste Siting Commission. *LLRW Frontline*. Spring 1988/k.

New York State Low-Level Radioactive Waste Siting Commission. *LLRW Frontline*. Fall 1988/l.

New York State Low-Level Radioactive Waste Siting Commission. *Public Meeting Summary Report*. January 1989/a.

New York State Low-Level Radioactive Waste Siting Commission. *Executive Summary - Source Term Report: Low-Level Radioactive Waste Projections for New York*. July 1989/b.

New York State Low-Level Radioactive Waste Siting Commission. *Disposal Method Screening Report*. July 1989/c.

New York State Low-Level Radioactive Waste Siting Commission. *Report on Potential Site Identification*. September 1989/d.

New York State Low-Level Radioactive Waste Siting Commission. *LLRW Frontline*. Spring 1989/e.

New York State Low-Level Radioactive Waste Siting Commission. *LLRW Frontline*. Summer 1989/f.

New York State Low-Level Radioactive Waste Siting Commission. *LLRW Frontline*. Winter 1989/g.

New York State Low-Level Radioactive Waste Siting Commission. *Generic Site Characterization Plan*. December 1989/h.

New York State Low-Level Radioactive Waste Siting Commission. *Evaluation of the Feasibility of the Mined Repository Option*. April 1990/a.

New York State Low-Level Radioactive Waste Siting Commission. *A Summary Status Report of Disposal Method Development*. November 1990/b.

New York State Low-Level Radioactive Waste Siting Commission. *Audit Report*. February 1991/a.

New York State Low-Level Radioactive Waste Siting Commission. *Disposal Program*. July 1991/b.

New York State Low-Level Radioactive Waste Siting Commission. *Minutes*. October 1988/m.

New York State Low-Level Radioactive Waste Siting Commission. *Minutes*. May 1992/a.

New York State Low-Level Radioactive Waste Siting Commission. *Minutes*. June 1992/b.

New York State Low-Level Radioactive Waste Siting Commission. *Minutes*. July 1992/c.

New York State Low-Level Radioactive Waste Siting Commission. *Minutes*. August 1992/d.

New York State Low-Level Radioactive Waste Siting Commission. *Minutes*. September 1992/e.

New York State Low-Level Radioactive Waste Siting Commission. *Minutes*. October 1992/f.

New York State Low Level Waste Group (NYSLLWG). *Revised Draft Position Paper on Interim Storage Plan*. August 1989/a.

New York State Low Level Waste Group. *Position Paper on Task Force (NYSERDA) Revised Interim Storage Plan*. November 1989/b.

New York State Low Level Waste Group. *Nationwide Interim Storage Update*. March 1990.

New York State vs. United States et al. Supreme Court of the United States. 1992.

Novikov, V. *The Nuclear Legacy of the Former USSR*. Presentation at the International Workshop on Risk and Fairness. International Institute for Applied Systems Analysis. Laxenburg, Austria. June 1993.

Oberholzer-Gee, F. Personal Communication. January 2, 1993.

O'Connor, R. E., & Shaw, L. G. *Technological Imperatives and Public Policies: Federal and State Roles in Managing Radioactive and Chemical Wastes*. Presented at Midwest Political Science Association, Chicago. April 14, 1988.

O'Hare, M., Bacow, L. S., & Sanderson, D. *Facility Siting and Public Opposition*. New York: Van Nostrand. 1983.

Oswego County Legislature. *Legislative Resolution in Opposition to the New York State Designation of the Towns of Scribia and New Haven as a Potential Area for a State Low-Level Waste Disposal Facility*. February 9, 1989.

Parker, F. L., Kasperson, R. E., Andersson, T. L., & Parker, S. A. *Technical and Sociopolitical Issues in Radioactive Waste Disposal, 1986*, Volumes 1A and II. Stockholm: The Beijer Institute. 1987.

Quinn, R. E. *Beyond Rational Management*. San Francisco: Jossey-Bass. 1988.

Quinn, R. E., & Rohrbaugh, J. W. A Competing Values Approach to Organizational Analysis. *Public Productivity Review, 5*, 141-159. 1981.

Quinn, R. E., & Rohrbaugh, J. W. A Spatial Model of Effectiveness Criteria: Towards a Competing Values Approach to Organizational Effectiveness. *Management Science, 29*, 363-377. 1983.

Quinn, R. E., Rohrbaugh, J. W., & McGrath, M. R. Automated Decision Conferencing: How it Works. *Personnel, 62*, 49-55. 1985.

Resnikoff, M. *Living Without Landfills*. New York: Radioactive Waste Campaign. 1987.

Renn, O. *Equity in Siting Issues: The Aargau Participation Project.* Presentation at the International Workshop on Risk and Fairness. International Institute for Applied Systems Analysis. Laxenburg, Austria. June 1993.

Renn, O., Webler, T., & Wiedemann, P. (Eds.) *Novel Approaches to Environmental Conflict Resolution: A Systemic Evaluation.* Boston: Kluwer Academic Publishers. Forthcoming.

Ricaud, J. L., & Delaunay, H. *Waste Management in France: Operations at COGEMA's UP3 Reprocessing Plant and Other Key Events.* Proceedings of the Symposium on Waste Management. Tucson, Arizona. March 1992.

Ristoratore, M. Siting Toxic Waste Disposal Facilities in Canada and the United States: Problems and Prospects. *Policy Studies Journal,* **14,** 140-148. 1985.

Rohrbaugh, J. W. Assessing the Effectiveness of Expert Teams. In J. L. Mumpower, L. D. Phillips, O. Renn, & V. R. R. Uppuluri (Eds.), *Expert Judgment and Expert Systems.* Berlin: Springer Verlag. 1987.

Rosa, E. A., Dunlap, R. E., & Kraft, M. E. Prospects for Public Acceptance of a High-Level Nuclear Waste Repository in the United States: Summary and Implications. In R. E. Dunlap, M. E. Kraft, & E. A. Rosa (Eds.), *The Public and Nuclear Waste: Citizens' Views of Repository Siting.* Durham: Duke University Press. 1993.

Schuiling, E., & Anderson, G. *Citizen Participation in Selecting a Disposal Site for California's Low-Level Radioactive Waste.* Manuscript. 1991.

Siting Process Task Force on Low-Level Radioactive Waste Disposal. *Opting for Co-Operation.* December 1987.

Siting Task Force on Low-Level Radioactive Waste Management. *Draft Impact Management Guidelines.* March 1989.

Siting Task Force on Low-Level Radioactive Waste Management. *Site Elimination Criteria.* Third Draft. January 1990/a.

Siting Task Force on Low-Level Radioactive Waste Management. *Opting for Co-Operation: A Process in Action. The First Phases.* Synopsis report. August 1990/b.

Siting Task Force on Low-Level Radioactive Waste Management. *Background Document: Chronology, Comparative Costs and Employment Opportunities.* December 1990/c.

Siting Task Force on Low-Level Radioactive Waste Management. *Progress Report to the Minister of Energy, Mines, and Resources.* January 1993.

Slovic, P. *Perception of Risk From Radiation.* Proceedings of the National Council on Radiation Protection and Measurements, Washington, DC. April 1989.

Slovic, P., Fischhoff, B., & Lichtenstein, S. The Psychometric Study of Risk Perception. In V. T. Covello, J. Menkes, & J. Mumpower

(Eds.), *Risk Evaluation and Management* (pp. 3-24). New York: Plenum. 1986.

Snyder, P., Monaco, C., & Harrington, J. *The Siting Commission Evaluation of Geology and Mines.* Cortland County LLRW Office. November 1991.

Solomon, B. D., Shelley, F. M., Pasqualetti, M. J., & Murauskas, G. T. Radioactive Waste Management Policies in Seven Industrialized Democracies. *Geoforum*, **18**, 415-431. 1987.

Stanley, R., Griepentrog, J., & Shealy, H. *Letter to Governor M. Cuomo.* October 1, 1990.

Tanaka, E. *Letter to the State of California Department of Health Services.* September 27, 1991.

Teigen, K. H., Brun, W., & Slovic, P. Societal Risks as Seen by a Norwegian Public. *Journal of Behavioral Decision Making*, **1**. 1988.

Texas, State of. *Low-Level Radioactive Waste Disposal Authority Act.* Chapter 273. 1981.

Texas, State of, Low-Level Radioactive Waste Disposal Authority (TLLRWDA). *Considerations for Site Selection.* November 1982.

Texas, State of, Department of Health, Bureau of Radiation Control (TDOH). Texas Regulations for the Control of Radiation (TRCR) Part 45, *Licensing Requirements for Near-Surface Land Disposal of Radioactive Waste.* October 1983.

Texas, State of, Low-Level Radioactive Waste Disposal Authority. *Texas Low-Level Radioactive Waste Disposal Facility Conceptual Design.* December 1983/a.

Texas, State of, Low-Level Radioactive Waste Disposal Authority. *Texas Low-Level Radioactive Waste Storage Facility.* December 1983/b.

Texas, State of, Low-Level Radioactive Waste Disposal Authority. *Economic Impact of Radioactive Waste Disposal Sites in Texas.* February 1984/a.

Texas, State of, Low-Level Radioactive Waste Disposal Authority. *An Assessment of the Social and Special Effects of Siting a Low-Level Radioactive Waste Disposal Facility in Texas.* February 1984/b.

Texas, State of, Low-Level Radioactive Waste Disposal Authority. *Siting a Low-Level Radioactive Waste Facility in Texas: Potential Impacts and Intergovernmental Relations.* March 1984/c.

Texas, State of, Low-Level Radioactive Waste Disposal Authority. *An Analysis of Public Opinion on Low-Level Radioactive Waste Disposal in Selected Texas Counties.* April 1984/d.

Texas, State of, Low-Level Radioactive Waste Disposal Authority. *Biennial Report on the Operations of the Texas Low-Level Radioactive Waste Disposal Authority and the Status of Interstate Compacts and*

Agreements Related to Low-Level Radioactive Waste Disposal. November 1984/e.

Texas, State of. *Low-Level Radioactive Waste Disposal Authority Act-Amendments.* Chapter 692. 1985.

Texas, State of, Low-Level Radioactive Waste Disposal Authority. *Siting a Low-Level Radioactive Waste Disposal Facility in Texas: Local Government Participation, Mitigation, Compensation, Incentives, and Operator Standards.* January 1985.

Texas, State of, Low-Level Radioactive Waste Disposal Authority and Texas Department of Health, Bureau of Radiation Control. *Institutional Low-Level Radioactive Waste in Texas: Current Waste Management Practices and Recommendations.* March 1985.

Texas, State of, Low-Level Radioactive Waste Disposal Authority. *An Assessment of the Social and Special Effects of Siting a Low-Level Radioactive Waste Disposal Facility in Texas.* April 1986/a.

Texas, State of, Low-Level Radioactive Waste Disposal Authority. *An Analysis of Public Opinion on Low-Level Radioactive Waste Disposal.* October 1986/b.

Texas, State of. *Low-Level Radioactive Waste Disposal Authority Act-Amendments.* Chapter 22. 1987.

Texas, State of. *Low-Level Radioactive Waste Disposal Authority Act-Amendments.* Chapters 76 and 945. 1987.

Texas, State of, Low-Level Radioactive Waste Disposal Authority. *Low-Level Radioactive Waste Disposal Facilities: Conceptual Designs and Assessments, 1-5.* February 1987.

Texas, State of, Low-Level Radioactive Waste Disposal Authority. *Report to the Public on the Site Selection Process and Other Analyses.* December 1990.

Texas, State of. *Low-Level Radioactive Waste Disposal Authority Act-Amendments.* Chapter 402. 1991.

Texas, State of. *Low-Level Radioactive Waste Disposal Authority Act-Amendments.* Chapter 583. 1991.

Texas, State of, Low-Level Radioactive Waste Disposal Authority. *Program Update.* July 1992.

Thibaut, J., & Walker, R. *Procedural Justice: A Psychological Analysis.* Hillsdale, New Jersey: Erlbaum. 1975.

United States Department of Energy (DOE). *The 1979 State-by-State Assessment of Low-Level Radioactive Wastes Received at Commercial Disposal Sites.* National Low-Level Radioactive Waste Management Program. 1980.

United States Department of Energy. *The 1980 State-by-State Assessment of Low-Level Radioactive Wastes Received at Commercial Disposal Sites.* National Low-Level Radioactive Waste Management Program. 1981.

United States Department of Energy. *The 1981 State-by-State Assessment of Low-Level Radioactive Wastes Received at Commercial Disposal Sites*. National Low-Level Radioactive Waste Management Program. 1982.

United States Department of Energy. *The 1982 State-by-State Assessment of Low-Level Radioactive Wastes Received at Commercial Disposal Sites*. National Low-Level Radioactive Waste Management Program. 1983.

United States Department of Energy. *The 1983 State-by-State Assessment of Low-Level Radioactive Wastes Received at Commercial Disposal Sites*. National Low-Level Radioactive Waste Management Program. 1984/a.

United States Department of Energy. *Nuclear Waste Management and the Use of the Sea*. National Advisory Committee on Oceans and Atmosphere (NACOA). DOE/LLW-69T. 1984/b.

United States Department of Energy. *The 1984 State-by-State Assessment of Low-Level Radioactive Wastes Received at Commercial Disposal Sites*. National Low-Level Radioactive Waste Management Program. 1985.

United States Department of Energy. *The 1985 State-by-State Assessment of Low-Level Radioactive Wastes Received at Commercial Disposal Sites*. National Low-Level Radioactive Waste Management Program. 1986/a.

United States Department of Energy. *A Multiattribute Utility Analysis of Sites Nominated for Characterization for the first Radioactive-Waste Repository - A Decision-Aiding Methodology*. DOE/RW-0074. 1986/b.

United States Department of Energy. *The 1986 State-by-State Assessment of Low-Level Radioactive Wastes Received at Commercial Disposal Sites*. National Low-Level Radioactive Waste Management Program. 1987.

United States Department of Energy. *The 1987 State-by-State Assessment of Low-Level Radioactive Wastes Received at Commercial Disposal Sites*. National Low-Level Radioactive Waste Management Program. 1988/a.

United States Department of Energy. *Data Base for 1988: Spent Fuel and Radioactive Waste Inventories, Projections, and Characteristics*. DOE/RW-0006, Rev. 4. Washington, DC: DOE. 1988/b.

United States Department of Energy. *The 1988 State-by-State Assessment of Low-Level Radioactive Wastes Received at Commercial Disposal Sites*. National Low-Level Radioactive Waste Management Program. 1989.

United States Department of Energy. *The 1989 State-by-State Assessment of Low-Level Radioactive Wastes Received at Commercial Disposal Sites*. National Low-Level Radioactive Waste Management Program. 1990.

United States Department of Energy. *The 1990 State-by-State Assessment of Low-Level Radioactive Wastes Received at Commercial Disposal Sites*. National Low-Level Radioactive Waste Management Program. 1991.

United States District Court Northern District of New York. *Complaint of The State of New York, The County of Allegany, N.Y., and The County of Cortland, N.Y.* February 1990.

United States General Accounting Office (GAO). *Report to the Honorable J. James Exon, U.S. Senate: Extensive Process to Site Low-Level Waste Disposal Facility in Nebraska.* July 1991.

United States General Accounting Office. *Nuclear Waste: New York's Adherence to Site Selection Procedures is Unclear.* August 1992.

United States Nuclear Regulatory Commission. *Rules and Regulations, Title 10, Chapter 1. Code of Federal Regulations - Energy. Part 61.* December 1982.

US Ecology. *Report to the People of Nebraska: Background Information Low-Level Radioactive Waste.* May 1988.

US Ecology. *Progress: Information for Nebraskans.* Spring 1990.

US Ecology. *Progress: Information About the Low-level Radioactive Waste Management Project.* 1991.

van Dorp, F., Diebold, P., Kappeler, S., Sattel, G., Tripet, P., Resele, G., & Wiborgh, M. *Application of a Waste Disposal Site Investigation Programme Based on Safety Analysis.* Proceedings of an International Symposium on the Siting, Design and Construction of Underground Repositories for Radioactive Wastes. Hannover, West Germany. March 1986.

van der Pligt, J. *Nuclear Energy and the Public.* Oxford: Blackwell. United Kingdom. In press.

Vari, A., & Farago, K. From Open Debate to Position War: Siting a Radioactive Waste Repository in Hungary. *Waste Management,* **11**. 1991.

Vari, A., Kemp, R., & Mumpower, J. L. Public Concerns about LLRW Facility Siting: A Comparative Study. *Journal of Cross-Cultural Psychology,* **22**, 83-102. 1991.

von Winterfeldt, D., & Edwards, W. Patterns of Conflict About Risky Technology. *Risk Analysis,* **4**, 55-68. 1984.

Williams, E. A., & Massa, A. *Siting of Major Facilities: A Practical Approach.* New York: McGraw-Hill. 1983.

White, I. *Letter from the New York State Energy Research and Development Authority to Jerome Griepentrog, Department of Human Resources, State of Nevada.* April 23, 1990/a.

White, I. *Letter from the New York State Energy Research and Development Authority to Roger Stanley, Department of Ecology, State of Washington.* April 23, 1990/b.

White, I. *Letter from the New York State Energy Research and Development Authority to Heyward Shealy, Department of Health and Environmental Control, State of South Carolina.* April 23, 1990/c.

White, I. *Letter from the New York State Energy Research and Development Authority to Jerome Griepentrog, Department of Human Resources, State of Nevada.* December 6, 1990/d.

White, I. L., & Spath, J. P.   How Are States Setting Their Sites? *Environment*, **26**, 17-20, 36-42. 1984.

Woodruff, P. A., Friedman, T., Bentley, C., Hayden, T., Quackenbush, C., Brulte, J., Lempert, T., Chandler, C., Gotch, M., Bates, T., Farr, S., Margolin, B., & Connelly, L.   *Letter of State Legislators to Molly Coye*, Director, California Department of Health Services.   October 21, 1991.

Working Group on State Assistance to Localities Affected by the Siting of a Low-Level Radioactive Waste.   *Proposal on Local Assistance to Communities Affected by the Siting of LLRW Disposal Facilities.* October 27, 1988.

APPENDIX

Table A.1

REQUIREMENTS FOR PERMANENT DISPOSAL SITES, AS SPECIFIED BY 6 NYCRR PART 382 REGULATIONS

(a) General
  (aa) the site must not be located at the Western New York Nuclear Service Center in West Valley;
  (ab) the site must be capable of being licensed and permitted under all applicable state and federal laws;
  (ac) the primary emphasis in disposal site suitability must be given to isolation of wastes;
  (ad) the site and the associated hydrology, geologic formations, and groundwater flow systems affecting facility performance must be capable of being defensibly characterized, modeled, analyzed, and monitored; and
  (ae) the site must include a buffer zone within which waste will not be disposed of, treated, or handled in any way, except of the transportation of the waste from the site boundary to the disposal facility.

(b) Surface Waters and Hydrology
  (ba) disposal units must not be located downstream of a dam or other man-made or natural structure; and
  (bb) the site must be located where the potential for adverse effects on surface water quality is minimized.

(c) Mineral and Exploitable Resources
  (ca) the site must not be located in an area where past or present exploration or exploitation of natural resource exists.

(d) Geology and Soils
  (da) the present and projected geologic and hydrogeologic setting of the site must be compatible with the waste disposal method; and
  (db) the site geology and soils must have properties that will retard potential movement of radionuclides.

(e) Groundwater and Hydrogeology
  (ea) the disposal facility must be located in a hydrogeologic regime that minimizes, to the extent practicable, the potential movement of radionuclides; and

Table A.1 Continued

(eb) the site must be located where the potential for adverse effects on groundwater quality is minimized and where groundwater use will not be impaired.

(f)   Seismic Risk

(fa) areas must be avoided where tectonic processes such as faulting, folding, seismic activity, or volcanism may occur.

(g)   Population, Land Use, and Ownership

(ga) the site must not be located where present and projected effects from population growth and other developments would have the potential to adversely affect the facility's ability to meet performance objectives;

(gb) the site must be located on property to which the State of New York can obtain title;

(gc) the site must be located to minimize, to the extent practicable, potential population dose based on consideration of population density in the vicinity of the site;

(gd) no part of the site may be located within towns or cities having at least 1000 individuals per square mile;

(ge) the site must be of sufficient size to ensure performance objectives and must be a minimum of 500 acres in surface areas; and

(gf) the site must not be located on any lands or reservations of Indian tribes, or nations or lands not subject to, or controlled by, the laws and regulations of the State of New York, including but not limited to, lands owned by the federal government.

(h)   Traffic and Transportation

(ha) the site must be located so that any new access route to be constructed between existing local highways or railroads and the site will be compatible with present land usage; and

(hb) the condition of existing highways and rail lines likely to be used as routes to the site must be adequate, or capable of being made adequate.

(i)   Cultural, Recreational, and Natural Resources

(ia) the site must not be located on lands protected by Article 14 of the State Constitution;

(ib) the site must not be located on any part of the National Wildlife Refuge System;

(ic) the site must not be located in state wildlife management areas, game refuges, fish hatcheries, or game farms;

Table A.1 Continued

(id) the site must not be located on properties acquired with the assistance of federal funding pursuant to the Dingell-Johnson Fish Restoration Act;

(ie) the site must not be located on migratory bird reservations;

(if) the site must not be located in reforestation areas;

(ig) the site must not be located on properties listed, nominated, or eligible for listing in the National Register of Historic Places;

(ih) the site must not be located on land dedicated tot he State Nature and Historical Reserve;

(ii) the site must not be located on properties subject tot he State Wild, Scenic and Recreational River System;

(ij) the site must not be located on land that is part of the National Wilderness Preservation system;

(ik) the site must not be located on wetlands acquired or restored with state moneys;

(il) the site must not be located within any national, state, county and municipal park systems established as of July 31, 1987;

(im) the site must be located to minimize, to the extent practicable, adverse impacts on agricultural lands; the site must not be located on lands in active agricultural use classified as mineral soil groups 1-4;

(in) the site must be located to minimize, to the extent practicable, adverse impacts on the environment; and

(io) the site must not be located in the critical habitat of any endangered or threatened species.

Additional criteria for any site used for above-ground or below-ground disposal methods:

(a) Surface Waters and Hydrology

(aa) the disposal site must not be placed in a 100-year floodplain, coastal hazard area, or wetland;

(ab) the disposal site must not be located in freshwater wetlands, a tidal wetland, or coastal erosion hazard area;

(ac) the surface contours and drainage systems of the disposal site must ensure that the site is well-drained and free of areas of flooding or frequent ponding; and

(ad) the disposal units must not be located in areas where surface runoff could adversely affect the facility's ability to meet the performance objectives.

Table A.1 Continued

(b) <u>Groundwater and Hydrology</u>
  (ba) the hydrogeologic unit used for disposal must not discharge groundwater tot he surface within the disposal site;
  (bb) the disposal site must have sufficient depth to the water table that groundwater intrusion, perennial or otherwise, will not contact disposal units; and
  (bc) the site must not be located immediately above the Long Island aquifer; any primary public water supply aquifer; or a principal aquifer designated by DEC.

(c) <u>Geology and Soils</u>
  (ca) areas must be avoided where surface and subsurface geologic conditions or processes such as mass wasting, subsidence, erosion, solutioning, slumping, landsliding, or weathering occur with such frequency and extent to significantly affect the long-term stability of the site.

  Additional criteria for any site used for underground mined repository:

(a) <u>Surface Water and Hydrology</u>
  (aa) the portion of the disposal facility located on the surface must not be placed in a 100-year floodplain, coastal hazard area, or wetland.

(b) <u>Mined and Exploitable Resources</u>
  (ba) areas must be avoided where existing mined opening and/or boreholes could adversely affect the facilities ability to meet performance objectives.

(c) <u>Groundwater and Hydrology</u>
  (ca) the geologic unit in which the mined repository is excavated must have low permeability;
  (cb) the present and expected regional groundwater flow must be capable of being defensibly characterized, modeled, and analyzed;
  (cc) the regional groundwater flow system must provide a sufficiently long pathway and slow transit time from the repository to the discharge area;
  (cd) aquifers between the ground surface and the disposal units, which could flood the disposal units, must be avoided; and

Table A.1 Continued

        (ce) the disposal unit must not be located in an area where subsurface hydrogeologic processes, such as dissolution, occur with such frequency and extent to significantly affect the ability of the facility to meet performance objectives.

(d)   Geology

        (da) the site must have geologic formations and hydrogeologic features which permit safe construction, operation, and closure of a disposal facility;

        (db) the geologic unit in which the repository is located must have adequate thickness and lateral extent to contain the disposal units;

        (dc) the geologic unit in which the repository is located must not contain features, such as faults, fractures, and other discontinuities, the presence of which would prevent the facility from meeting performance objectives; and

        (dd) the site must be located in an area where subsidence and its impacts are minimized to the extent practicable.

Source: NYSDEC, 1987/b

Table A.2

CRITERIA FOR ASSESSING POTENTIAL SITES FOR LLRW DISPOSAL, AS PROPOSED IN THE DRAFT SITE SELECTION

(1)   Geology
    (i)   Prefer areas that contain undeformed to very gently folded stratigraphic units devoid of lenses, dikes, fractures, faults, or other discontinuities that will increase the difficulty of adequately characterizing the site and predicting performance;
    (ii)   prefer areas that have lower predicted maximum horizontal ground acceleration;
    (iii) prefer areas that are distant from a known or suspected Quaternary fault;
    (iv)   prefer areas that do not demonstrate significant past or active subsurface dissolution; and
    (v)   prefer host geologic units that can accommodate the thickness and lateral space required for the facility and a buffer zone.

(2)   Natural Resources
    (i)   Prefer areas that are distant from active or nearby abandoned mines;
    (ii)   prefer areas that do not have viable economic mineral resource potential; and
    (iii) prefer areas that are distant from active or abandoned oil or gas fields, underground injection wells, underground gas storage areas, and areas of solution mining.

(3)   Groundwater Hydrology
    (i)   Prefer areas that are distant from primary or principal aquifers, hydrogeologic units that fit the definition of principal aquifers, and well head areas that directly recharge public water supply wells and public surface water supplies;
    (ii)   prefer areas that are distant from surface water features that are sustained by groundwater discharge, such as lakes, reservoirs, perennial and ephemeral streams, springs, seeps, swamps, marshes, and bogs; and
    (iii) prefer stratigraphic units that have a high clay or silt content.

(4)   Surface Water Hydrology
    (i)   Prefer areas that are away from wetland as defined in Executive Order 11988, freshwater or tidal wetlands designated pursuant to ECL Articles 24 or 34, or areas that are under federal jurisdiction as wetlands;

Table A.2 Continued

- (ii) prefer sites that are well drained and free from areas of potential flooding or ponding from streams or runoff;
- (iii) prefer areas exhibiting no existing or potential erosional characteristics that could adversely affect waste containment;
- (iv) prefer areas where there is no potential for flooding from failure of upstream man-made impoundments; and
- (v) prefer areas that are located away from significant surface water resources in which water quality, as measured against water quality standards or impairment of best intended usage, could be adversely affected by site activities.

(5)   Meteorology
- (i) Prefer areas with relatively low annual amounts of precipitation that accumulate from low-intensity precipitation events;
- (ii) prefer sites distant from areas prone to severe weather phenomena such as heavy snowfall; and
- (iii) prefer areas that exhibit less than average depths of frost penetration.

(6)   Air Quality
- (i) Prefer areas where a significant PSD increment is available for consumption, especially for total suspended particulates and inhalable particulates.

(7)   Ecology
- (i) Prefer areas that are distant from critical habitat of terrestrial or aquatic species that are threatened or endangered, or prime fish/game recreational areas.

(8)   Radiology
- (i) Prefer areas that are as far as practical from any existing facility or source that generates and/or releases radioactive materials that may mask the environmental monitoring program.

(9)   Demography
- (i) Prefer areas that exhibit low population densities;
- (ii) prefer areas that are away from highly populated areas (greater than 2,500 persons); and
- (iii) prefer areas that are far away from concentrations of nonresident population.

(10) Land Use
- (i) Prefer areas that are distant from federal protected lands;
- (ii) prefer areas that are distant from lands protected by New York State;
- (iii) prefer areas that are distant from Indian lands;

Table A.2 Continued

(iv) prefer government-owned lands that have not been excluded from consideration; and

(v) prefer areas where development of significant local population growth appear not to be imminent.

(11) Transportation

(i) Prefer areas near the interstate highway system, other major roadways, existing rail lines, or waterways;

(ii) prefer areas served by routes which are not currently congested;

(iii) prefer areas that utilize routes which do not require new construction or structural improvements to handle waste shipments;

(iv) prefer areas where access is available by more than one mode;

(v) prefer areas that are close to major waste generators;

(vi) prefer areas for which the access routes exhibit low accident rates, as measured by actual numbers of accidents, resulting property damage, and lives lost; and

(vii) prefer areas served by routes that do not pass through populated places upon leaving the major transportation system (e.g., interstate highway system).

(12) Socioeconomics

(i) Prefer areas where there is a sufficient work force available with the required labor skills to construct and operate the facility;

(ii) prefer areas with an adequate or excess housing stock; and

(iii) prefer areas where municipal services (e.g., sewer, water, schools, emergency services) have not reached capacity and the disposal facility will not create a disproportionate demand on services.

(13) Cultural Resources/Aesthetics

(i) Prefer sites which do not contain other known or suspected archaeological, cultural, or historical resources;

(ii) prefer areas where the disposal area will not degrade the present viewshed; and

(iii) prefer areas which do not have any sensitive noise receptors (e.g., houses) in close proximity.

Source: NYSLLRWSC, 1988/h

Table A.3

SUMMARY STATISTICS FOR ALL SITE SELECTION WEIGHTS

Criteria	Above-Ground Technologies		Below-Ground Technologies		Mine Technologies	
	Mean	Standard Deviation	Mean	Standard Deviation	Mean	Standard Deviation
RADIOLOGY						
Existing Facilities	45.6	35.4	42.5	34.9	49.5	39.5
DEMOGRAPHY						
Population Density	46.5	44.1	42.9	46.3	41.9	57.1
Highly Populated Areas	32.1	22.5	27.5	25.1	22.3	27.7
Nonresident Population	15.8	8.5	13.8	9.3	13.7	13.3
LAND USE						
Federally Protected Land	9.5	10.2	8.1	9.6	7.0	7.7
New York Protected Land	11.0	10.0	10.1	9.8	9.8	9.5
Indian Lands	12.6	8.6	11.1	8.5	11.0	9.2
Non-exclusive Government Lands	21.8	30.4	19.0	27.8	13.4	22.5
Development/Growth	19.7	17.9	17.4	17.3	16.7	15.3
TRANSPORTATION						
Major Routes	11.8	12.6	11.6	12.7	11.5	12.0
Uncongested Routes	7.5	8.0	7.3	8.0	6.1	7.0
Existing Routes	5.0	7.4	4.8	7.2	3.8	6.8

Table A.3 Continued

Multi-modal Access	4.1	5.3	4.0	5.3	6.3	11.4
Proximity to Generators	25.8	55.0	18.6	30.2	20.6	34.8
Accident Rates	16.0	22.0	15.2	21.6	18.9	22.5
Avoid Populated Areas	9.5	11.0	9.3	11.0	8.9	11.7
SOCIOECONOMICS						
Labor Force	8.9	10.0	8.7	10.3	9.9	12.4
Housing	10.0	10.9	10.1	10.6	5.8	7.1
Municipal Service	13.3	9.7	13.8	12.1	10.0	12.2
CULTURAL RESOURCES						
Archeological	23.0	17.1	21.8	16.7	26.0	26.5
Viewshed	13.3	9.2	10.0	9.9	7.2	8.0
Noise Receptors	11.8	9.1	9.2	8.8	6.4	7.3
GEOLOGY						
Stratigraphic Complexity	36.5	34.3	44.9	37.2	81.8	52.4
Ground Acceleration	20.8	18.4	20.6	19.5	19.8	23.1
Quaternary Faults	15.5	19.5	15.4	18.4	11.2	15.2
Subsurface Dissolution	35.0	30.0	32.6	17.5	53.7	33.4
Thickness and Lateral Extent	35.0	31.0	32.3	22.1	64.5	35.7
NATURAL RESOURCES						
Mines	13.1	13.2	24.9	23.2	37.4	23.9
Mineral Resources	16.2	15.3	23.1	17.4	41.0	24.2

Table A.3 Continued

Oil and Gas	16.4	14.8	24.5	14.0	49.5	28.9
GROUNDWATER HYDROLOGY						
Aquifers	41.9	35.2	66.6	37.8	101.4	36.7
Surface Water	37.8	39.3	50.6	25.5	63.0	28.0
Clay/Silt Content	32.3	23.3	53.8	37.1		
SURFACE WATER						
Wetlands	32.4	16.7	35.9	19.6		
Drainage/Ponding	38.2	29.0	35.7	20.8		
Erosion	38.0	30.1	34.3	22.9		
Upstream Impoundments	30.5	22.5	23.0	13.2	42.1	31.2
Resources	28.7	20.7	28.7	19.0	40.2	28.7
METEOROLOGY						
Precipitation	20.2	21.1	17.0	14.0		
Severe Weather	31.2	19.8	12.0	11.0	14.1	15.7
Frost Penetration	15.4	15.8	7.1	7.5		
AIR QUALITY						
PSD Increment	32.4	31.0	28.1	26.2	15.0	18.1
ECOLOGY						
Critical Habitat/Prime Areas	57.8	29.0	52.0	29.7	38.6	26.0

Source: NYSLLRWSC 1988/e

Table A.4

RECOMMENDED BASE CASE WEIGHTING FACTORS FOR SITING CRITERIA: ABOVEGROUND/BELOWGROUND TECHNOLOGY

Criteria	Workshop (8/88)	Staff (9/88)	Recommended Base Case Weight Set
Geology			
1. Stratigraphic Complexity	40.7	46.8	45
2. Ground Acceleration	20.7	23.7	20
-- Quaternary Faults	15.5	Not Weighted	--
3. Subsurface Dissolution	33.8	30.8	35
5. Thickness/Lateral Extent	33.7	30.4	35
Subtotal	144.4	131.7	135
Natural Resources			
7. Mines	19.0	20.5	20
8. Mineral Resources	19.7	25.7	20
9. Oil and Gas	20.5	26.4	20
Subtotal	59.2	72.6	60
Groundwater Hydrology			
12. Aquifers	54.3	60.3	55
13. Surfacewater	44.2	40.0	40
14. Clay/Silt Content	43.1	34.1	40
Subtotal	141.6	134.4	135
Surface Water Hydrology			
18. Wetlands	34.1	21.0	30
19. Drainage/Ponding	37.0	21.1	30
20. Erosion	36.2	30.0	35
21. Upstream Impoundments	26.8	20.1	25
22. Resources	28.7	30.4	30
Subtotal	162.8	122.6	150

Table A.4 Continued

Criteria	Workshop (8/88)	Staff (9/88)	Recommended Base Case Weight Set
Meteorology			
23. Precipitation	18.6	18.1	20
25. Severc Weather	21.6	22.4	20
-- Frost Penetration	11.3	Not Weighted	--
24. Snowfall	Not Weighted	15.1	15
Subtotal	51.5	55.6	55
Air Quality			
27. PSD Increment	30.3	26.1	30
Ecology			
29. Critical Habitat	54.9	55.5	55
Radiology			
30. Radiology	44.1	21.3	20
31. Performance Objectives	Not Weighted	22.1	20
Subtotal	44.1	43.4	40
Demographic Patterns			
33. Population Density	44.7	46.9	45
34. Highly Populated Areas	29.8	47.3	40
35. Nonresident Population	14.8	26.7	20
Subtotal	89.3	120.9	115
Land Use			
37. Federal Protected Lands	8.8	10.1	10
39. State Protected Areas	10.6	10.7	10
42. Indian lands	11.9	12.8	10
43. Non-exclusive Public Land	20.4	24.6	20
45. Development/Growth	18.6	18.4	20
Subtotal	70.3	76.6	70

Table A.4 Continued

Criteria	Workshop (8/88)	Staff (9/88)	Recommended Base Case Weight Set
Transportation			
47. Major Routes	11.7	8.7	10
48. Uncongested Routes	7.4	7.1	10
49. Existing Routes	4.9	6.9	5
50. Multi-model access	4.1	9.2	5
51. Proximity to generators	22.2	21.5	20
52. Accident Rates	15.6	9.9	15
53. Avoid Populated Places	9.4	11.3	10
Subtotal	75.3	74.6	75
Socioeconomics/Community Services			
54. Labor Force	8.8	13.6	10
55. Housing	10.1	8.8	10
56. Municipal Services	13.6	16.2	15
Subtotal	32.5	38.6	35
Cultural Resources			
59. Archeological/Historic	22.4	17.3	20
60. Viewshed	11.7	18.0	15
61. Noise	10.5	13.2	10
Subtotal	44.6	48.5	45
TOTALS	1,000.8	1,001.1	990

Source: NYSLLRWSC, 1988/i

Table A.5

FINAL CRITERIA FOR SELECTING SITES FOR A LLRW DISPOSAL FACILITY, INCLUDING MEASUREMENT SCALES

(a)   Geology

(aa) Prefer areas that contain uniformly distributed soils, sediments, and/or rock that are relatively undeformed and devoid of fractures, faults, and other discontinuities that may influence predictions of performance

       1 - hydrogeologic modeling could be complex;
       3 - hydrogeological modeling would not be complex;
       5 - hydrogeological modeling would be relatively straightforward;

(ab) prefer areas of lower seismic hazard based on predicted maximum horizontal ground acceleration

       1 - greater than 20 percent g;
       3 - 10 to 20 percent g;
       5 - less than 10 percent g;

(ac) prefer areas that do not demonstrate significant past or active subsurface dissolution

       1 - sites exhibits sinkholes, caverns, or under ground streams;
       3 - no surface or subsurface evidence of dissolution is present;
       5 - site is not subject to dissolution by groundwater;

(ad) exclude all abandoned mines and all geologic units that are less than or equal to 30 meters below ground surface (exclusionary criteria);

(ae) prefer host geologic units that can accommodate the vertical and horizontal space required for all disposal facilities and any necessary buffer zone

       1 - less than 10 m or less than 300 acres;
       5 - greater than 10 m and greater than 300 acres.

(b)   Natural Resources

(ba) Exclude all reforestation areas or parts thereof established pursuant to Article 9, Title 5 of the ECL (exclusionary criteria);

Table A.5 Continued

    (bb) prefer areas that are distant from active or nearby abandoned mines
        1 - less than or equal to 1/2 mile;
        3 - greater than 1/2 mile to 1 mile;
        5 - greater than 1 mile;

    (bc) prefer areas that do not have likely economic mineral resource potential
        1 - site has past, current, or future mineral or energy resource exploration;
        3 - site has undergone past resource exploration and no evidence was found for economic reserves;
        5 - site is situated within a geologic setting that exhibits no economic resource potential;

    (bd) prefer existing mines and nearby areas that do not have economic mineral or energy resource potential
        1 - existing mine and area within 1 mile has identified economic reserves or is undergoing exploration;
        3 - existing mine has no future economic potential and area within 1 mile has minimal economic potential;
        5 - existing mine has no future economic potential and area within 1 mile has no future economic potential;

    (be) prefer areas that are distant from active or abandoned oil or gas fields, underground injection wells, underground gas storage areas, and areas of solution mining
        1 - less than or equal to 1/2 mile;
        3 - greater than 1/2 mile to 1 mile;
        5 - greater than 1 mile.

(c)   <u>Groundwater Hydrology</u>
    (ca) Exclude all areas immediately above the Long Island aquifer, any primary public water supply aquifer, or a principal aquifer designated by DEC (exclusionary criteria);
    (cb) prefer areas that are distant from primary or principal aquifers, hydrogeological units that fit the definition of principals aquifers, and well head areas for community water supply systems

Table A.5 Continued

          1 - less than or equal to 1/2 mile;

          3 - greater than 1/2 mile to 1 mile;

          5 - greater than 1 mile;

(cc) prefer areas that are distant from significant surface water features that are likely to be sustained by groundwater discharge through the site

          1 - less than or equal to 1/2 mile;

          3 - greater than 1/2 mile to 1 mile;

          5 - greater than 1 mile;

(cd) prefer unconsolidated stratigraphic units that restrict groundwater flow and retard radionuclide movement

          1 - site contains unconsolidated units that have minor or no clay content;

          3 - site contains unconsolidated units that have moderate clay content;

          5 - site contains thick unconsolidated units that have high clay content.

(d)   Surface Water Hydrology

    (da) Exclude all significant perennial surface water bodies (exclusionary criteria);

    (db) exclude 100-year floodplains; coastal erosion hazard areas; or flood hazard areas (exclusionary criteria);

    (dc) exclude wetlands, freshwater wetlands, and tidal wetlands (exclusionary criteria);

    (dd) prefer areas that are distant from wetlands, freshwater wetlands, tidal wetlands, areas under federal jurisdiction as wetlands, areas with hydric soils, areas incorporated in the National Wetlands Inventory Classification, and extensive areas of less than 2 percent slope

          1 - less than or equal to 1/4 mile;

          3 - greater than 1/4 mile to 1 mile;

          5 - greater than 1 mile);

    (de) prefer sites that are well drained and free from areas of potential ephemeral flooding or ponding from streams and runoff

          1 - site exhibits poor run-off characteristics;

          2 - site exhibits run-off characteristics that indicate it is subject to ponding or flooding during storms;

Table A.5 Continued

                  3 - site exhibits relatively good runoff
characteristics;

                  4 - site exhibits good runoff characteristics;

                  5 - site is located in areas not subject to flooding or
ponding even during storms;

(df) prefer sites exhibiting no existing or potential erosional
characteristics that could adversely affect waste containment

                  1 - site exhibits significant erosional features;

                  3 - site does not exhibit significant erosional
features but may be subject to erosion during
storm events;

                  5 - site is not subject to significant erosion from
surface runoff;

(dg) prefer areas where there is no potential for flooding from failure
of upstream man-made impoundments

                  1 - site is located where flooding resulting from
failure of impoundments could adversely affect
facility performance;

                  3 - site is located where flooding resulting from
failure of impoundments is unlikely or could
adversely affect facility performance;

                  5 - site is located where no man-made
impoundments are present;

(dh) prefer areas where location of a disposal facility is unlikely to
impair the best usage of surface waters, as measured against
water quality standards

                  1 - location of facility is likely to affect surface
water quality;

                  3 - location of facility could potentially affect
surface water quality;

                  5 - location of facility is unlikely to affect surface
water quality.

(e)   Meteorology and Climatology

(ea) Prefer areas with relatively low annual amounts of precipitation

                  1 - greater than or equal to 50 inches/year;

                  3 - 40 to less than 50 inches/year;

                  5 - less than 40 inches/year;

(eb) prefer areas that are not prone to chronic severe weather
phenomena such as heavy snowfall

                  1 - greater than or equal to 100 inches annually;

Table A.5 Continued

                    3 - 50 to less than 100 inches annually;

                    5 - less than 50 inches annually;

    (ec) prefer sites that are not prone to frequent incidents of severe weather

                    1 - high frequency of incidents;

                    3 - medium frequency of incidents;

                    5 - low frequency of incidents.

(f)   Air Quality

    (fa) Exclude areas that are within pollutant nonattainment areas (exclusionary criteria);

    (fb) prefer areas where a significant PSD increment is available for consumption, especially for total suspended particulate and inhalable particulate

                    1 - less than 20 percent increment available;

                    5 - greater than 20 percent increment available.

(g)   Ecology

    (ga) Exclude all areas where threatened or endangered species or designated critical habitats for such terrestrial or aquatic species or species of special concern are present (exclusionary criteria);

    (gb) prefer areas that are distant from critical habitat of terrestrial or aquatic species that are threatened or endangered, and areas inhabited by other important species

                    1 - adjacent to site;

                    3 - in close proximity to site;

                    5 - distant from site.

(h)   Incompatible Nearby Activities

    (ha) Prefer areas that are as far as practical from any existing facility or natural source that could generate and/or release radioactive materials that may mask the environmental monitoring program

                    1 - known releases with high potential to mask monitoring;

                    3 - potential releases that might mask monitoring;

                    5 - releases unlikely to mask monitoring;

    (hb) prefer areas that are distant from nearby activities that could interfere with the ability of the facility to meet performance objectives

                    1 - nearby activities with high potential for interference;

Table A.5 Continued

        3 - nearby activities with potential for interference;

        5 - nearby activities unlikely to interfere.

(i) Demographic Patterns

  (ia) Exclude all villages, towns, cities, or unincorporated places that have an average population density of more than 1,000 persons per square mile (exclusionary criteria);

  (ib) prefer areas that exhibit low population densities

        1 - 700 to 1,000 persons per square mile;

        2 - 400 to 699 persons per square mile;

        3 - 150 to 399 persons per square mile;

        4 - 50 to 149 persons per square mile;

        5 - 0 to 49 persons per square mile;

  (ic) prefer areas that are away from highly populated areas (greater than 2,500 persons)

        1 - less than 1 mile;

        2 - 1-3 miles;

        3 - 4-7 miles;

        4 - 8-10 miles;

        5 - greater than 10 miles;

  (id) prefer areas that are distant from concentrations of nonresident population

        1 - significant presence;

        3 - some presence;

        5 - none.

(j) Land Use

  (ja) Exclude all lands protected by the Federal government, including components of The National Wilderness Preserve System, National Wildlife Refuge System, National Wild and Scenic Rivers System, Migratory bird reservations, fish restoration areas, and the National Park System (exclusionary criterion);

  (jb) prefer areas that are distant from lands protected by the Federal government

        1 - less than or equal to 1/2 mile;

        3 - greater than 1/2 to 1 mile;

        5 - greater than 1 mile;

  (jc) exclude all lands protected by New York State, including components of New York State Wild, Scenic, and Recreational

Table A.5 Continued

Rivers System, Fish restoration areas, the State Park System, Adirondack Park, Catskill Park, Municipal parks established as of December 31, 1987, Wildlife Management areas, game refuges, game farms, fish hatcheries, and boat launches (exclusionary criterion);

(jd) prefer areas that are distant from lands protected by New York State
> 1 - less than or equal to 1/2 mile;
> 3 - greater than 1/2 to 1 mile;
> 5 - greater than 1 mile;

(je) exclude all real property that cannot acquired for facility use by the State of New York (exclusionary criterion);

(jf) exclude all Indian reservations and lands under jurisdiction of Indian nations (exclusionary criterion);

(jg) prefer areas that are distant from Indian lands
> 1 - less than or equal to 1/2 mile;
> 3 - greater than 1/2 to 1 mile;
> 5 - greater than 1 mile;

(jh) prefer government-owned lands that have not been excluded from consideration
> 1 - private ownership;
> 3 - federal/locally owned land;
> 5 - state-owned land;

(ji) exclude all lands in mineral soil groups 1-4, as designated by the New York State Land Classification System, that are in active agricultural production (exclusionary criteria);

(jj) prefer areas where development or significant local population increases will not be imminent
> 1 - high likelihood;
> 3 - some potential;
> 5 - highly unlikely;

(jk) exclude all lands within the boundaries of the Western New York Nuclear Service Center in West Valley, New York (exclusionary criteria).

(k) Transportation

(ka) Prefer areas near the interstate highway system, other major (multilane, limited access) roadways, or existing rail lines
> 1 - greater than 10 miles;
> 3 - 5 to 10 miles;
> 5 - less than 5 miles;

Table A.5 Continued

(kb) prefer sites served by routes which are not currently congested

      1 - greater than 1.2 volume-to-capacity (V/C) ratio;

      2 - 1.0 to 1.2 V/C ration;

      3 - .7 to 1.0 V/C ratio;

      4 - .4 to .7 V/C ratio;

      5 - less than .4 V/C ratio;

(kc) prefer sites that utilize routes which do not require new construction or structural improvements to handle waste shipments

      1 - significant new construction or upgrading;

      3 - moderate new construction or upgrading;

      5 - no improvements necessary;

(kd) prefer sites where access is available by more than one mode

      1 - rail only,

      3 - highway only;

      5 - highway and rail;

(ke) prefer areas that are close to major waste generators

      1 - significantly greater than weighted mean distance;

      3 - approximately equal to weighted mean distance;

      5 - significantly less than weighted mean distance;

(kf) prefer sites for which the access routes exhibit low accident rates, as measured by actual numbers of accidents, resulting property damage, and lives lost

      1 - not used for this criterion

      2 - measures of accident rates significantly exceed statewide averages;

      3 - measures of accident rates approximate the statewide average;

      4 - measures of accident rates are significantly less than statewide averages;

      5 - not used for this criterion

(kg) prefer sites served by routes that do not pass through incorporated areas upon leaving the major transportation system (e.g., interstate highway system)

      1 - encounter populated places;

      3 - bypassed populated places;

      5 - no populated places.

Table A.5 Continued

(l)   Socioeconomics/Community Services
       (la) Prefer sites where there is a sufficient work force available with
             the required labor skills to construct and operate the disposal
             facility
                            1 - no local labor force;
                            3 - some in-migration needed;
                            5 - no in-migration needed;
       (lb) prefer areas with an adequate or excess housing stock (in terms
             of quality, number, and affordability) within a reasonable
             commuting distance or time
                            1 - no excess present;
                            3 - some additional required;
                            5 - no additional required;
       (lc) prefer sites, within reasonable commuting distance or time,
             where municipal services (e.g., sewer, water, schools,
             emergency services) have not reached capacity and the disposal
             facility will not create a disproportionate demand on services
                            1 - large required investment to meet demand;
                            3 - moderate required investment to meet demand;
                            5 - no investment to meet demand.
(m)   Cultural Resources/Aesthetics
       (ma) Exclude all areas that are listed, nominated, or eligible for
             listing in the National or State Registers of Historic Places
             (exclusionary criteria);
       (mb) exclude all lands that have been dedicated to or acquired for the
             purpose of being dedicated to the State Nature and Historical
             Preserve.
       (mc) prefer areas that do not contain other archaeological, cultural,
             or historical resources
                            1 - known resources on site;
                            3 - no known resources;
                            5 - no resources within the site based on previous
                            surveys;
       (md) prefer areas where the disposal facility will not degrade the
             present scenery or viewshed
                            1 - impacts to views cannot be mitigated;
                            3 - no impacts to views, with mitigation;
                            5 - no impact to views;

Table A.5 Continued

    (me) prefer areas which do not have any people or animals in close
           proximity that may be sensitive to noise
                    1 - many within 1 mile;
                    3 - few within 1 mile;
                    5 - none within 1 mile

Source: NYSLLRWSC, 1988/h

Table A.6

CRITERIA AND WEIGHTING FACTORS USED IN THE SECOND STEP OF THE CANDIDATE AREAS IDENTIFICATION PROCESS

Exclusionary

(1) depth of geologic unit;
(2) principal aquifers;
(3) significant surface waters;
(4) air pollutant nonattainment areas;
(5) the Western New York Nuclear Service Center;
(6) the National Wild and Scenic River System; and
(7) state wildlife management areas and Wild, Scenic, and Recreational
       River Systems.

Preference	Weights
(8) earthquake acceleration and ground movement	53
(9) thickness, areal extent, and composition of geologic unit;	200
(10) distance from oil and gas fields;	53
(11) distance from aquifers;	147
(12) distance from groundwater discharge zones	100
(13) surface water quality;	80
(14) chronic severe weather - snowfall;	53
(15) precipitation;	53
(16) low population density;	120
(17) distance from Federal protected lands;	27
(18) distance from State protected lands;	27
(19) distance from Indian lands; and	27
(20) proximity to waste generators.	100
	1,000

Source: NYSLLRWSC, 1988/i

Table A.7

THIRTY CANDIDATE AREAS AND THEIR PERFORMANCE
SCORES

(1)	Cape Vincent, Jefferson County	4360-4788*
(2)	Orleans/LeRay, Jefferson County	4348-4788
(3)	Carisle/Charleston/Glen, Schoharie and Montgomery Counties	4228-4788
(4)	Lisbon/DePayster/Oswagatchies, St. Lawrence County	4242-4576
(5)	Altona/Ellenburg, Clinton County	4416-4576
(6)	Summit, Schoharie County	4494-4788
(7)	Coventry, Chenango County	4788
(8)	McDonough/Preston/Pharsalia/Plymouth/Symrna /Linckloen/DeRuyter/Georgetown/Lebanon,Chenango and Madison Counties	3934-4894
(9)	Fort Ann/Hartford, Washington County	4682
(10)	Crawford/Montgomery/Shawangunk/Shawangunk Hill, Orange and Ulster Counties	3828-4668
(11)	Fenner/Sullivan/Lenox/Verona, Madison and Oneida Counties	3880-4788
(12)	Berkshire/Anticoke, Tioga and Broome Counties	3762-4894
(13)	Solon/Taylor/Freetown/Cincinnatus/Marathon/Willet, Cortland County	3868-4788
(14)	Parish/Amoby/West Monore, Oswego County	4508-4682
(15)	Barton/Tioga, Tioga County	4374-4880
(16)	Scriba/New Haven, Oswego County	4457-4668
(17)	Sennett/Owasco/Niles, Cayuga County	4176-4894
(18)	Lodi/Covert/Ovid/Ulysses, Seneca and Tompkins County	4400-4894
(19)	Throop/Aurelius/Springport/Fleming/Scopio/Ledyard/ Venic/Genoa/Lansing, Cayuga and Tompkins Counties	3894-4894
(20)	Greenville/Wawayanda, Orange County	4402-4562
(21)	Sterling/Victory/Cato/Ira, Cayuga County	3894-4788
(22)	West Almond/Ward/Andover, Allegany County	4470-4682
(23)	Granger/Allen/Candeado, Allegany County	4136-4682
(24)	Torrey/Milo/Barrington/Wayne, Yates and Steuben Counties	3828-4788
(25)	Gorham/Seneca/Potter, Ontario and Yates Counties	3882-4788

Table A.7 Continued

(26)	Canandaigua/Bristol, Ontario County	4242-4668
(27)	Somerset/Hartland/Royalton/Niagara County	4402-4456
(28)	Prattsburg, Steuben County	4494-4788
(29)	Caton, Steuben County	4894
(30)	Jasper/Camerson/Howard, Steuben County	4282-4788

*(ranges of scores indicate different levels of performance across the area)

Source: NYSLLRWSC, 1988/i

Table A.8

SUMMARY OF THE KEY PROVISIONS OF NEW YORK STATE DEPARTMENT OF ENVIRONMENTAL CONSERVATION REGULATIONS, 6 NYCRR, PART 382

Performance objectives:

(a) Protection of the general population from releases of radioactivity;
(b) Protection of individuals from inadvertent intrusion;
(c) Protection of individuals during operations;
(d) Stability of the disposal site after closure; and
(e) Protection of public health and the environment.

General criteria:

(a) Disposal of mixed hazardous and LLRW wastes must meet the applicable requirements of Article 27 of Environmental Conservation Law and the applicable regulations;
(b) Construction methods and techniques must not adversely affect the facility's ability to meet the performance objectives;
(c) Activities performed during the institutional control period not to be relied upon more than 100 years;
(d) The disposal units must function in concert with other features of the site to ensure that performance objectives are met: Complement natural characteristics of the site, compatibility with methods of closure, long-term isolation, and avoidance of need for continuing active maintenance;
(e) The performance of the disposal units must be capable of being characterized, modeled, analyzed, and evaluated;
(f) Recovery or retrieval of waste from the disposal units must be feasible during the institutional control period; and
(g) The facility should possess long-term structural stability (100 years for Class A wastes, 300 years for Class B wastes, and 500 years for Class C wastes).

Additional criteria for above-ground and below-ground disposal:

(a) The design must ensure that the geochemical interaction between the disposal unit containment structures and the soils will not impair the ability to meet the performance objectives;

Table A.8 Continued

(b)   The disposal units must be placed on stable foundations such that settlement of the structures does not impair the ability to meet the performance objectives;

(c)   The disposal units must be designed to maintain their structural integrity and containment performance under major natural phenomena (e.g., floods, earthquakes, etc.);

(d)   The disposal units must be designed to minimize leakage of water into the units;

(e)   The disposal units must be provided with reliable, redundant monitoring systems;

(f)   The disposal units must function in concert with the other features of the disposal site to ensure that the performance objectives are met: Water infiltration must be minimized, percolating or surface water must be directed away, site must resist degradation by surface geologic processes and biotic activity, surface water drainage must not result in erosion, contact of waste with water and snow must be minimized; and

(g)   The disposal units must be designed to minimize the amount of radioactive material that would be released into the environment should water infiltrate the disposal units.

Criteria for above-ground disposal only:

(a)   The disposal units must be designed to resist to inadvertent intrusion (100 years for Class A wastes, 300 years for Class B wastes, and 500 years for Class C wastes);

(b)   The disposal units must maintain their structural integrity and containment performance under normal environmental conditions; and,

(c)   The disposal units must facilitate remedial actions during the active institutional control period.

Criteria for below-ground diposal only:

(a)   Below-ground disposal units must resist inadvertent intrusion for a period of at least 500 years, if they contain Class C wastes, or must provide for the placement of Class C waste so that the top of the waste is a minimum of 5 meters below the top surface of the cover material.

Table A.8 Continued

Criteria for underground mined repositories:

(a) Waste disposal operations must not interfere with mining operations;

(b) Damage to the surrounding rock must be minimized;

(c) All shafts, access tunnels, or adits and all bores in proximity must be sealed;

(d) The repository must be located at a depth below the ground surface of greater than 30 meters;

(e) Subsidence of the ground surface must be minimized;

(f) The contact of water with waste should be minimized; and

(g) The repository must comply with federal and New York State regulations on mining operations.

Table A.9

**DRAFT CRITERIA FOR SELECTING METHODS FOR DISPOSAL
OF LLRW**

(a) Public Safety
  (aa) Dose to public: Keep the total radiation exposures to the
      general public from the facility as low as reasonably achievable
      and within regulatory limits.
(b) Worker Safety
  (ba) Dose to workers: Keep the total radiation exposures to
      workers at the facility as low as reasonably achievable.
  (bb) Industrial safety: Minimize the number of lost-time
      nonradioactive injuries during construction, waste handling
      activities, and closure activities at the facility.
(c) Inadvertent Intruder Safety
  (ca) Dose to inadvertent intruders: Prevent an intruder from being
      exposed to or inadvertently causing the release of radionuclides
      from the facility after closure.
  (cb) Resistance to inadvertent intruders: Prefer methods that are
      resistant to inadvertent intruders.
(d) Long-Term Effectiveness of Barrier Systems
  (da) Barrier effectiveness: The increased number and type of
      barriers used in a facility design will provide some assurance
      of protection against infiltration of water, degradation by
      surface geologic processes, and biotic activity.
  (db) Structural stability: Separate Class A, B, and C waste forms
      and isolate waste for 100, 300, and 500 years, respectively.
  (dc) Water leakage management: Features to minimize the impact
      of water intrusion into the disposal units should be provided in
      the facility design.
  (dd) Construction materials: Construction materials of the disposal
      units should be durable, long-lived, and not prone to
      deterioration when exposed to adverse soil or weather
      conditions.
(e) Impact on Local Community
  (ea) Economic impacts: Disposal facility operation will be
      economically beneficial to the community by providing
      employment and local business opportunities.
  (eb) Aesthetics: The disposal facility should be aesthetically
      attractive or at least aesthetically neutral.

Table A.9 Continued

(f) Economics and Costs
    (fa) Life-cycle costs: The total life-cycle costs should be as low as practicable consistent with meeting the performance objectives for LLRW disposal.
    (fb) Operational cost sensitivity: Each disposal method will be evaluated to determine the impact that variations in the waste stream will have on unit burial costs. If generators implement waste volume reduction programs, the facilities may be oversized and able to operate for longer periods. Because of the differences in fixed and variable costs, this may cause a significant change in unit disposal costs. Conversely, if the generator waste streams become larger and/or more difficult to dispose, facility operating life and/or disposal costs could be adversely affected.

(g) Licensability
    (ga) Licensing experience: Apply for and obtain license from NYS DEC within time limitations set by federal law. Facility must be in operation by 1 January 1993.
    (gb) Operational experience: Evaluate methods based on operating experience.

(h) Method Practicality
    (ha) Surveillance and monitoring: Monitoring to detect increases above background radioactivity or pollutant concentrations in the vicinity of the disposal site caused by failure of disposal units.
    (hb) Impact of construction: The disposal unit construction methods and techniques must not adversely affect the land disposal facility's ability to meet the performance objectives.

(i) Nonradiological Environmental Impacts
    (ia) Emission control: 6 NYCRR Part 382.15 requires that construction, operation, closure, and the post-closure conditions of the land disposal facility must comply with all applicable laws and regulations including but not limited to environmental, labor, and public health laws, and regulations.
    (ib) Land use: Minimize land area required.

(j) Operational Flexibility
    (ja) Accept all forms/containers: The low-level radioactive waste facility should be able to accept all waste forms and containers that are described in NRC Branch Technical Position on waste forms.

Table A.9 Continued

    (jb) Waste handling required: Minimize the number and complexity of waste handling operations.

    (jc) Impact of construction on operations: Prefer construction activities that are simple and less likely to impact waste handling.

    (jd) Obtainable sources of supply: The disposal method should be based on easily obtainable materials, services, supplies, and equipment.

(k)   Efficiency of Mitigative Measures

    (ka) Remedial action: The disposal method should be amenable to remedial actions needed to repair the disposal unit and clean up contamination released in case of failure.

    (kb) Retrieval and recovery: The disposal method should have the ability to retrieve or recover the waste. "Recovery" means removing from a disposal unit waste that has been permanently disposed in a land disposal facility. "Retrieval" means recovery of waste in an intact container.

Source: NYSLLRWSC, 1988/b

Table A.10

SUMMARY STATISTICS FOR ALL METHOD SELECTION
WEIGHTS

Criteria	Mean	Standard Deviation
Dose to Public	152.4	50.1
Dose to Workers	83.7	43.5
Industrial Safety	38.2	19.4
Dose to Inadvertent Intruders	41.4	27.0
Resistance to Inadvertant Intruders	47.6	25.0
Barrier Effectiveness	48.7	26.5
Structural Stability	37.5	27.1
Water Leakage Management	50.6	29.7
Construction Materials	37.9	24.0
Economic Impact on Local Community	20.2	24.3
Aesthetics	12.6	14.9
Life Cycle Costs	31.8	27.8
Operational Cost Sensitivity	46.0	63.8
Licensing Experience	18.6	20.2
Operational Experience	21.3	20.0
Surveillance and Monitoring	48.2	23.2
Impact of Construction	26.0	18.3
Emission Control	24.0	21.0
Land Use	16.0	16.6
Accept All Forms/Containers	20.3	16.4
Waste Handling Required	26.9	16.9
Impact of Construction on Operations	13.1	12.3
Obtainable Sources of Supply	9.6	8.5
Remedial Action	63.7	35.8
Retrieval and Recovery	63.7	47.9

Source: NYSLCRWSC, 1988/e

Table A.11

RECOMMENDED BASE CASE WEIGHTING FACTORS FOR
METHOD SELECTION CRITERIA

	Criteria	Workshop	Staff	Recommended Base Weight Set
1.	Dose to Public	152.4	141.1	147
2.	Dose to Workers	83.7	69.6	77
3.	Industrial Safety	38.2	45.5	42
4.	Dose to Inadvertent Intruders	41.4	27.0	34
5.	Resistance to Inadvertent Intruders	47.6	37.5	42
6.	Barrier Effectiveness	48.7	34.8	42
7.	Structural Stability	37.5	32.3	35
8.	Water Leakage Management	50.6	33.3	42
9.	Confidence in Construction Materials	37.9	39.2	38
10.	Economic Impact on Community	20.2	47.6	34
11.	Aesthetics of Facility	12.6	30.5	22
12.	Total Life Cycle Costs	31.8	73.6	53
13.	Operational Cost Sensitivity	46.0	33.5	40
14.	Ease of Licensing	39.9	74.6	57
15.	Environmental Surveillance and Monitoring	48.2	48.6	48
16.	Impact of Required Construction Methods	26.0	22.5	24
17.	Emission Control	24.0	27.5	26
18.	Land Use Requirements	16.0	23.1	20
19.	Ability to Accept All Waste Forms and Containers	20.3	35.4*	28
20.	Impact of Activities on Facility Operations	40.0	21.7*	31
21.	Easily Obtainable Sources of Supply	9.6	9.1*	9

Table A.11 Continued

22.	Ease and Effectiveness of Remedial Action	63.7	50.6	57
23.	Ease and Effectiveness of Retrieval and Recovery	63.7	41.5	52
	TOTAL		1,000	1,000

*These data were modified to reflect the addition of criteria 21 after the staff weighting exercise. The weights for operational flexibility were redistributed based on ratios of the Workshop weights.

Source: NYSLLRWSC, 1990/a

TABLE A.12

FINAL CRITERIA FOR SELECTING METHODS FOR DISPOSAL OF LLRW, INCLUDING MEASUREMENT SCALES

(a) Public Safety

(a) Dose to public: Exclude all disposal methods with projected annual doses to the public in excess of NYCRR Part 382.11 limits at any time during the life-time of the facility. Prefer disposal methods that keep the normally expected radiation exposures to any individual of the general public from the facility as low as reasonably achievable and within regulatory limits.

1 - highest dose to the public
5 - lowest dose to the public

(b) Worker Safety

(ba) Dose to workers: Exclude all disposal methods that result in annual doses to workers in excess of 6NYCRR Part 382.13 and 12NYCRR 38.21. Prefer disposal methods that keep the aggregate radiation exposures to all exposed workers at the facility as low as reasonably achievable.

1 - highest dose to workers
5 - lowest dose to workers

(bb) Industrial safety: Prefer disposal methods that minimize the number of lost-time nonradioactive injuries over the life of the facility, from design through institutional control.

1 - maximum number of lost-time injuries
5 - smallest number of lost-time injuries

(c) Inadvertent Intruder Safety

(ca) Dose to inadvertent intruders: Exclude all disposal methods that, by their design, cannot limit the dose to inadvertent intruders to less than 500 millirems/yr. Prefer disposal methods that minimize the exposure to inadvertent intruders.

1 - 500-400 millirems/yr
2 - 400-300 millirems/yr
3 - 300-200 millirems/yr
4 - 200-100 millirems/yr
5 - less than 100 millirems/yr

(cb) Resistance to inadvertent intruders: Exclude all disposal methods that cannot be designed to resist inadvertent intruders for a period of 100 years for Class A, 300 years for Class B, and 500 years for Class C waste. Prefer methods that are inherently resistant to inadvertent intruders for the appropriate periods of time for each Class of waste.

Table A.12 Continued

                         1 - does not inherently resist inadvertent intruders and
                                 requires the addition of significant engineered
                                 intruder barriers
                         5 - inherently resistant to inadvertent intruders
        (d) <u>Long-Term Effectiveness of Barrier Systems</u>
                         (da) Barrier effectiveness: Exclude all disposal methods
that provide no resistance to the infiltration of water into the disposal units.
Prefer disposal methods that provide the most assurance of protection against
infiltration of water.
                         1 - barriers are least likely to prevent the infiltration of
                                 water,
                         5 - inherently resistant to the infiltration of water
                         (db) Structural stability: Exclude all disposal methods that
are not expected to remain structurally stable for 100 years for Class A
waste, 300 years for Class B waste, and 500 years for Class C waste. Prefer
disposal methods that provide the greatest confidence in the long-term
structural stability of the disposal units.
                         1 - requires extensive provisions to promote high
                                 confidence in the required long-term stability and
                                 containment of waste,
                         5 - requires few provisions to promote high confidence in
                                 the required long-term stability and containment of
                                 waste
                         (dc) Water leakage management: Exclude all disposal
methods that cannot be designed to include a drainage system or other
mechanism to minimize the contact time between the waste and water that
may leak into the disposal unit. Prefer disposal methods that minimize the
impact of water that may intrude into the disposal units.
                         1 - requires extensive provisions to promote high
                                 confidence in the ability to reliably remove water
                                 from the disposal unit,
                         5 - requires few provisions to promote high confidence in
                                 the ability to reliably remove water from the
                                 disposal unit
                         (dd) Construction materials: Exclude all disposal methods
that depend on construction materials that are quickly degraded by exposure
to surface geologic processes and biotic activity. Prefer disposal methods
that use construction materials historically known to be durable.
                         1 - critical dependence on unproven materials,

Table A.12 Continued

2 - substantial protection required from unproven materials,
3 - moderate degradation of performance results from
     unproven materials,
4 - minimal dependence on unproven materials,
5 - construction materials known to be durable

(e) Impact on Local Community

(ea) Economic impact on community: Prefer disposal
methods that provide the most acceptable economic effects to the affected
local communities (measures will be defined after input has been received
from the affected local communities).

(eb) Aesthetics of facility: Prefer disposal methods that are
most aesthetically acceptable to the affected local communities (measures
will be defined after input has been received from the affected local
communities).

(f) Economics and Costs

(fa) Total life-cycle costs: Prefer disposal methods that
maximize cost-effectiveness over the life of the facility, from design through
institutional control.

1 - highest life-cycle cost
5 - lowest life-cycle cost

(fb) Operational cost sensitivity: Prefer disposal methods
that show the least increase in operating cost with respect to variations in
quantity of waste.

1 - highest cost increase per unit volume
5 - lowest cost increase per unit volume

(g) Licensability

(ga) Ease of licensing: Exclude all disposal methods that
cannot meet licensing requirements, that are specifically prohibited, or that
are not covered by the existing regulations. Prefer disposal methods that
have been reviewed generically by the NRC and that have received favorable
comments.

1 - major technical concerns cited, or not reviewed, by
     NRC
5 - reviewed and recommended by NRC

(h) Method Practicality

(ha) Environmental surveillance and monitoring: Exclude
all disposal methods that cannot be provided with reliable, redundant
monitoring systems to detect the failure of the disposal unit. Prefer disposal
methods that can most easily be provided with reliable and

Table A.12 Continued

effective surveillance and monitoring system for early detection of failure of
disposal units.
>	1 - least reliable and least effective
>	5 - most reliable and most effective
>	(hb) Impact of required construction methods: Exclude all
disposal methods and techniques that would result in the failure of the facility
to meet the performance objectives of 6NYCRR 382, Subpart C. Prefer
disposal methods that use construction methods and techniques that are least
likely to adversely affect the land disposal facility's ability to meet the
performance objectives.
>	1 - high likelihood of damaging construction activity
>	5 - no likelihood of damaging construction activity

(i) Nonradiological Environmental Impacts
>	(ia) Emission control: Exclude all disposal methods that
would exceed regulatory limits for environmental emissions during
construction, operation, closure, or the post-closure period.
>	1 - maximum quantity of environmental emissions
>	5 - minimum quantity of environmental emissions
>	(ib) Land use requirements: Prefer disposal methods that
minimize the land area required for the disposal units consistent with an
adequate buffer zone.
>	1 - largest land requirement for disposal units
>	5 - smallest land requirement for disposal units

(j) Operational Flexibility
>	(ja) Ability to accept all waste forms and containers:
Prefer disposal methods that can readily accept a variety of waste forms and
containers.
>	1 - can handle only a single type of standardized container
>	and/or waste form
>	5 - can handle a variety of waste forms and types of LLRW
>	containers)
>	(jb) Impact of activities on facility operations: Prefer
disposal methods with required activities that are simple and less likely to
impede operations.
>	1 - required activities are complex and subject to long
>	delays
>	5 - required activities are of minimal difficulty
>	(jc) Easily obtainable sources of supply: Prefer disposal
methods that require materials, services, and equipment that are easily
obtainable through multiple suppliers.

Table A.12 Continued

             1 - disposal method is dependent on unusual materials,
                      services, equipment and/or a sole source supplier
             5 - disposal method uses easily obtainable sources of supply

(k) Efficiency of Mitigative Measures

        (ka) Ease and effectiveness of remedial action: Exclude all disposal methods for which remedial action is impossible. Prefer disposal methods for which effective remedial actions are easy to perform and inexpensive.

             1 - effective remedial action is difficult to perform, and
                      expensive
             5 - effective remedial action is easy to perform with low,
                      reasonable costs

        (kb) Ease and effectiveness of retrieval and recovery: Exclude all disposal methods that do not permit the retrieval or the recovery of waste from the disposal unit. Prefer disposal methods that permit the easiest access to waste in the disposal unit for the purpose of retrieval or recovery.

             1 - difficult access to the waste for retrieval/recovery
             5 - easy access to the waste for retrieval/recovery

Source: NYSLLRWSC, 1988/g

Table A.13

BASE CASE RANKING OF DISPOSAL METHODS

Class A Disposal Method Type	Total Weighted Score	Class B/C Disposal Method Type	Total Weighted Score
Surf. Mound MDU	3,594	Shall. Exc. Vault	3,301
Shall. Exc. Vault	3,543	Shall. Exc. MDU	3,221
Surf. Mound Vault	3,528	Surf Mound MDU	3,202
Shall. Exc. MDU	3,498	Shall. Exc. Liner	3,151
Surf. Only MDU	3,468	Shall. Exc. Mono.	3,112
Surf. Only Vault	3,439	MDU in a Tunnel	3,018
Shall. Exc. Mono.	3,354	Surf. Mound Vault	3,003
Shall. Exc. Liner	3,340	Vault in a Tunnel	2,996
Surf. Mound Liner	3,313	Monolith/Tunnel	2,943
Surf. Only Mono.	3,267	Lined Tunnel	2,862
Surf. Mound Mono.	3,264	Surf. Mound Liner	2,831
MDU in a Tunnel	3,242	MDU in a Mine	2,741
Vault in a Tunnel	3,178	Surf. Mound Mono.	2,740
Monolith/Tunnel	3,037	Monolith/Mine	2,685
Lined Tunnel	3,008	Lined Mine	2,669
MDU in a Mine	2,934	Surf. Only MDU	2,638
Lined Mine	2,746	Vault in a Mine	2,593
Mined Repository	2,741	Surf. Only Vault	2,587
Monolith/Mine	2,726	Surf. Only Mono.	2,457
Vault in a Mine	2,627	Mined Repository	2,323

Source: NYSLLRWSC 1989/c

INDEX

Acres International Limited, 171
Arkansas, State of, 109-111, 115
Arizona, State of, 67-70
Atomic Energy Act, 15-17
Atomic Energy Commission
    (AEC), 15, 17

Barnwell, South Carolina, 15, 18,
    22, 23, 131
Battelle Memorial Institute, 92,
    96, 98-99
Beatty, Nevada, 15, 18, 22, 71,
    84
Bechtel International, Inc., 124,
    128
below-regulatory concern, 22

California Radioactive Materials
    Management Forum, 68, 71-
    72, 83, 86-87
California, State of, 2, 13, 27,
    66-90, 113, 218-219, 222-228,
    230-232, 234, 236-239
    Department of Health
    Services, 67-68, 70-73, 82-86,
    88-89; Governor, 68-69, 84;
    Health and Welfare Agency,
    67; Inyo County, 72-73, 81;
    Legislature, 67-69, 84; Low-
    Level Radioactive Waste
    Advisory Committee, 71;
    Needles, City of, 67, 81, 83,
    87; Riverside County, 72-73,
    81; San Bernardino County,
    72-73, 81; Silurian Valley,
    81-82; Site Selection Citizens'
    Advisory Committee, 67, 72,
    75, 81-82, 89; State Land

Commission, 67, 84; Ward
    Valley, 67, 72, 81-84, 87-88
Canada, 1, 2, 4, 13, 160-176,
    215-221, 223, 225, 227, 230-
    232, 236-239
    Atomic Energy Canada
    Limited, 160; Atomic Energy
    Control Board (AECB), 161,
    163, 170-171, 231; Canadian
    Transport Commission, 163;
    Chalk River, 160-161, 168-
    169; Community Liaison
    Groups, 165-168; Deep River,
    161, 168-169; Department of
    Energy, Mines, and
    Resources, 162-163, 165, 170;
    Department of Transport, 163;
    Eldorado Resources, 160-164,
    168, 171, 173-175;
    Environment Canada, 163;
    Geraldton, 161, 168-169;
    Hope Township, 161, 163-
    164, 169; Hornepayne, 161,
    168-169; Lake Ontario, 160-
    161, 164; Low-Level
    Radioactive Waste
    Management Office, 163;
    Newcastle, 161, 163-165,
    169; Ontario, Province of,
    160-161, 163, 165-168; Port
    Granby, 160-162, 164-165;
    Port Hope, 160-163, 165,
    169, 175; Siting Process Task
    Force, 161-165, 171, 173-
    175; Siting Task Force, 161-
    163, 165-171; Welcome, 160,
    162, 165
Carter, President Jimmy, 18
Central Interstate LLRW
    Compact, 109-117, 126-127,
    129, 131-133, 226

303

Technology, Risk, and Society
An International Series in Risk Analysis

1. J.D. Bentkover, V.T. Covello and J. Mumpower (eds.): *Benefits Assessment*. The State of the Art. 1986   ISBN 90-277-2022-3

2. M.W. Merkhofer: *Decision Science and Social Risk Management.* A Comparative Evaluation of Cost-Benefit Analysis, Decision Analysis, and other Formal Decision-Aiding Approaches. 1987
                                                                 ISBN 90-277-2275-7

3. B.B. Johnson and V.T. Covello (eds.): *The Social and Cultural Construction of Risk.* Essays on Risk Selection and Perception. 1987
                                                                 ISBN 1-55608-033-6

4. R.E. Kasperson and P.J.M. Stallen (eds.): *Communicating Risks to the Public.* International Perspectives. 1990   ISBN 0-7923-0601-5

5. D.P. McCaffrey: *The Politics of Nuclear Power.* A History of the Shoreham Nuclear Power Plant. 1991   ISBN 0-7923-1035-7

6. M. Waterstone (ed.): *Risk and Society.* The Interaction of Science, Technology and Public Policy. 1991   ISBN 0-7923-1370-4

7. A. Vari and P. Tamas (eds.): *Environment and Democratic Transition.* Policy and Politics in Central and Eastern Europe. 1993
                                                                 ISBN 0-7923-2365-3

8. A. Vari, P. Reagan-Cirincione and J.L. Mumpower: *LLRW Disposal Facility Siting.* 1994   ISBN 0-7923-2743-8

Kluwer Academic Publishers – Dordrecht / Boston / London

The manufacturer's authorised representative in the EU is Springer
Nature Customer Service Centre GmbH, Europaplatz 3, 69115 Heidelberg,
Germany. If you have any concerns regarding our products, please
contact ProductSafety@springernature.com

Printed and bound by CPI Group (UK) Ltd, Croydon, CR0 4YY
29/04/2026
02099472-0002